P9-ECQ-355

KICKING THE CARBON HABIT

KICKING THE CARBON HABIT

Global Warming and the
Case for Renewable and Nuclear Energy

———

WILLIAM SWEET

Columbia University Press

NEW YORK

Columbia University Press
Publishers Since 1893
New York   Chichester, West Sussex

Copyright © 2006 by William Sweet
All rights reserved

Library of Congress Cataloging-in-Publication Data
Sweet, William.
Kicking the carbon habit : global warming and the case for renewable and
nuclear energy / William Sweet.
p.   cm.
Includes bibliographical references and index.
ISBN 0–231–13710–9 (cloth) — ISBN 0–231–51037–3 (electronic)
1. Global warming. [1. Renewable energy sources. 2. Power resources.] I. Title.
TJ808.S87   2006
333.79'4—dc22
2005035034

Columbia University Press books are
printed on permanent and durable acid-free paper.
This book is printed on paper with recycled content.
Printed in the United States of America

c 10 9 8 7 6 5 4

UNIVERSITY LIBRARY
UNIVERSITY OF ALBERTA

This one is for Gail.

Bruce Turton
Aug 2008

CONTENTS

PREFACE

This is not a book for those still trying to decide whether or not global warming is a problem worthy of attention, and still less is it a primer or textbook purporting to treat in an evenhanded fashion all significant aspects of the problem. For that, there are plenty of other places to go.[1] Rather, it's assumed that anybody picking the book up already senses that there's something to worry about and would like to know how serious the problem is and whether there's really anything we can do about it. The book is journalistic in that it's based mainly on interviews with leading experts and a close reading of some of what they've written, with a view to conveying the scientific and technical basis of the main concerns about global warming. It tells stories about people, and is written in hopes of being readable, interesting, and even at times somewhat entertaining. But it departs a little from the usual journalistic conventions by also making an unabashed argument, which in the end is solely the author's responsibility.

The main purpose of the book is to get across that there's been a revolution in climate science in the last fifty years—and that we are still a long way from acting adequately on the new knowledge acquired. Though policy is discussed in a global context, the focus is on the United States, because it represents about a quarter of the global problem, because of all the countries in the world, it is in the strongest position to reduce greenhouse gases sharply, and because Americans are the most likely readers.

The book does not necessarily have to be read straight through, start to finish. Readers, depending on their interests and proclivities, can begin with the second or third parts if they are keen to get to the science or the technology without delay, or even jump around from chapter to chapter. The first part of the book adds up the social and economic benefits and costs of making electricity from coal. The second describes how some pioneering scientists arrived at enormously important conclusions about earth's past and future climate. The third part discusses the low-carbon and zero-carbon technologies that can be deployed on a much larger scale right now to reduce greenhouse gas emissions, as well as some highly touted technologies that are in fact not ready for prime time.

KICKING THE CARBON HABIT

CHAPTER 1

The Case for Sharply Cutting U.S. Greenhouse Gas Emissions

IN THE EARLY and mid-1990s, a consensus developed among climate scientists that the world was warming alarmingly and that human activity was playing a role. That consensus, very cautiously expressed at first, was the basis of an international agreement for most advanced industrial countries to cut radiation-trapping greenhouse gas emissions to levels about 5 to 10 percent below those prevailing in 1990. Reducing emissions from combustion of critically needed oil, gas, and coal that much was, admittedly, an ambitious goal. But in light of the success the world had had in eliminating the chlorofluorocarbons that were eating up the earth's ozone shield, concerted international action to achieve the goal did not seem unattainable.

In the intervening years, many of the predicted effects of global warming have appeared, often sooner and more disturbingly than most scientists expected. As local habitats have changed in response to changing climates, species of plants and animals have migrated. In some places where they already live in cooler, high-lying areas, they have been climbing ever higher to stay in the only living conditions in which they can survive, but finding there's no place higher to go. Because of such situations, if present trends continue, massive species extinctions are predicted by this century's end. As the world's oceans absorb as much as half the carbon being emitted from human sources, waters are becoming markedly more acidic, threatening fragile corals and all manner of sea life. Meanwhile, Arctic ice has thinned by as much as 40 percent in the last three decades, and all the world's tropical and subtropical mountain glaciers are fast disappearing. Very soon, Hemingway's snows of Kilimanjaro will be no more.

As the world's oceans warm and expand, sea levels rise, threatening all low-lying coastal areas. At the same time, warmer waters provide fertile breeding grounds for cyclonic storms, which become fiercer.[1] Numerous Pacific islands; the great river deltas along China's east coast, Bangladesh, and nearby regions of India; the coastal areas of Holland, Belgium, and England; and the whole Gulf Coast of North America, from Florida to Mexico, all are severely at risk.

Hurricane Katrina, which devastated the Mississippi and Louisiana coasts in September 2005, flooding 80 percent of New Orleans, provided a vivid demonstration of what's at stake. Of course, no one event can be attributed to global warming, any more than a person's dying from lung cancer can be directly attributed to smoking or air pollution. The relationship is strictly statistical, a matter of probability. But there's little doubt that global warming makes events like Katrina more likely, and when they occur, the costs are staggering. Reconstruction and restoration of New Orleans and neighboring areas will cost the United States hundreds of billions of dollars.

According to one school of thought, we can't afford to take the measures required to curb global warming, and even if we do take them, others, by not going along, will cancel out the good effects of anything we do. This is the wrong way of looking at things. If your house is burning fiercely and neighboring houses are starting to catch fire too, of course you will try to round up your neighbors and get everybody to work together to put out the fires. But if for some reason there are laggards among them, you're not going to waste time trying to get them moving faster. You're going to get to work putting out your fire. It's not a moral issue. Working to protect your home is simply the normal thing to do.

It's not too much to characterize the situation facing the world as a global emergency. The most compelling computer projections, which will be described in detail in the sixth chapter of this book, show that as greenhouse gas levels double, triple, or even quadruple in this century, the wet areas of the world will get still wetter and the dry areas will become even more arid. In some of the great regions that we rely on most for our grains and foodstuffs, agriculture will become unsustainable, because the breadbaskets become too wet or too dry. Meanwhile, as critical infrastructure is put at risk, whole regional economic systems will be jeopardized.

Even worse, the prospect of a supraregional climate cataclysm cannot be ruled out. Paleoclimatologists studying records of earth's climate history preserved in high-altitude cores and ocean sediment—their work is described in detail in chapter 5—have found that whenever greenhouse gas levels were half the world's preindustrial levels, ice ages occurred. The changes in greenhouse gas levels now being induced by human activity are greater than the changes associated with the onset and termination of ice ages. As the levels are rising far above anything experienced and recorded in the last 700,000 years, we are entering wholly uncharted waters, where some kind of large-scale reorganization of the ocean-atmospheric system could occur. With present-day computational technology, we literally have no way of knowing what that might look like.

Given the enormity of the problem, it's easy to see why so many give in to despair, arguing that no effective action is possible. This point of view is found in the climate science community too, where it is not uncommon for those who take the pessimistic view to be also inclined to believe there's nothing to be done, as chapter 7 will show. But there are also those who take heart from progress already made—in reducing chlorofluorocarbons, for example—and who think greenhouse gases can be reined in if their sources are broken down into manageable elements.

In the spirit of thinking globally but acting locally, this book looks squarely at the United States, the world's biggest single source of greenhouse gases and the country best positioned to do something about them. The United States accounts for roughly a quarter of the world's greenhouse gases, about twice the share produced by Japan, Germany, Great Britain, France, and Italy combined.[2] While those five next-largest advanced industrial countries have ratified the international agreement requiring them to get their emissions below 1990 levels and have adopted national energy plans to fulfill that commitment, the United States has not. Since its emissions continue to rise, largely unconstrained by any positive policy, if and when this country finally decides to reverse course, it will have all the further to go.

Fortunately, the United States has a great deal of room for maneuver. It is by far the richest country in the world, in terms of both total income and per capita income. Taking the broadest view of technology, it also has by far the greatest technical resources. Most important, it uses energy much more extravagantly and carelessly than any other country, which means it can put its vast resources to work effectively to improve production and reduce consumption of energy very fast.

Anybody traveling in Europe in recent decades will have noticed all manner of conservation devices that are virtually unknown in the United States. It starts when you check into your hotel, where hallway lights are on timers and go off soon after you enter your room. The room itself has a key system that guarantees all electrical appliances are turned off when you leave, even if you'd prefer to keep your television and a light on for your return. Use of mass transit is ubiquitous. Gasoline prices are two or three times higher than in the United States, because of taxes, so cars are much smaller on average and people drive them fewer miles.

Of course the United States is a huge, sprawling country, with its own distinct culture, profoundly dependent on automotive transport. Realistically, it cannot be expected to suddenly transform itself into a completely different kind of place. While there no doubt is room for mass transportation to be expanded, Americans never will rely on subways, buses, and intercity trains the way people do in much more compact countries like Japan, Germany, or England. Nor will Americans readily give up their love affair with the automobile. Though some might imagine that we can come to grips with global warming just by persuading people to stop driving huge, gas-guzzling sport utility vehicles (SUVs), perhaps by closing the loophole in fuel-efficiency rules classifying SUVs as "light trucks" rather than cars, it would in fact take draconian measures to achieve automotive fuel savings adequate to the climate challenge facing us.

To get into step with what the other advanced industrial nations are doing to cut their greenhouse gas emissions, the United States needs to reduce its own by at least 25 percent, as soon as possible. The choices that must be made to accomplish that are explored in the third part of this book. To take the extreme case, discussed in chapter 10, if that entire reduction were to be accomplished in the automotive sector alone, Americans would have to accept gasoline prices that are two or three times higher than they are today—or accept rules that simply require them to buy and drive much more fuel-efficient vehicles. Though hybrid-electric cars have enormous promise, unless consumers are forced to buy them by higher fuel prices or legislative requirements, history shows that they will take advantage of standard vehicles' fuel efficiency to buy higher-performing cars or drive more miles, without achieving any net savings of fuel.

Altogether, the U.S. automotive sector accounts for roughly a third of the country's greenhouse gas emissions and electricity generation from coal for another third, with the rest coming from miscellaneous sources—mainly industrial processes. If only a fraction of the savings needed can come from motor vehicles, then the rest must come from converting the electricity sector to low or zero carbon generation and by conserving energy. Happily, as part 3 shows, this can be done, though some of the choices involved will not please everybody.

Per capita, citizens of the United States use almost twice as much energy as citizens of France, Germany, Italy, Japan, or the United Kingdom. In terms of output produced per amount of energy used, those other advanced industrial countries do about 50 percent better on average.[3] By the same token, the United States currently emits about twice as much greenhouse gas as Japan, Germany, and England, almost four times as much as Sweden, Switzerland, or France, 35 or 40 times as much as China, and nearly 100 times as much as India. Obviously, even allowing for profound differences in infrastructure, culture, and political economy, the United States can do a very great deal to use energy more efficiently and more carefully.[4] By putting the proper incentives into place—a suitable tax or cap on carbon emissions—we should be able to conserve enough energy to prevent electricity demand from growing for the rest of this century.

If at the same time a concerted push is made to sharply cut back on burning coal to generate electricity, a prompt 25 percent cut in carbon emissions ought to be achievable. There is tremendous potential for deploying wind turbines, as countries like Germany and Denmark have shown, even though not all their citizens like to see their landscapes dotted with huge steel towers. (The potential for expansion of solar and wind energy is discussed in chapter 9.) And there is still some room in the United States for replacing coal with natural gas, which is much cleaner and burns much more efficiently, producing only about half as much greenhouse gas per unit of electricity generated. But the United States is exhausting its domestic supplies of natural gas, imports cannot be boosted sharply without building controversial pipelines and liquefied natural gas terminals, and there are many competing uses of the valuable fuel—home heating, the production of chemicals, and even, if fuel cell–powered vehicles become prevalent in a couple of decades, motor vehicles. Therefore, it is argued in the eleventh chapter, some added reliance on nuclear power also will be necessary; some Americans may have to swallow their distaste for atomic energy and deal with its hazards, just as Europeans have had to accept some unpleasant trade-offs to cut greenhouse gas emissions.

The notion of sharply reducing reliance on coal is, admittedly, counter-intuitive. As radio and television advertisements sponsored by the U.S. coal and energy industries are constantly reminding us, the United States has enough coal reserves to meet all its energy needs for about 250 years. Those ads promise that with new technologies that will give us "clean coal," we can keep relying on coal—and perhaps rely on it even more—without unduly burdening the environment.

The important thing to grasp here is that in terms of carbon emissions, using present-day economically proven technology, there is no such thing as clean coal. Capturing carbon in emissions from coal-generating plants, the way sulfur dioxide and the nitrous oxides are trapped, and then finding a way to safely store huge quantities of the captured substance has not been demonstrated to be commercially via-ble, as the eighth chapter shows ("Breaking the Carbon Habit"). People in the industry may complain that wind energy or nuclear energy is too expensive, by comparison with coal, to warrant investment. But in fact, it's less expensive to replace a coal-fired electricity generator with a wind farm or a nuclear plant than it would be to capture and sequester the coal plant's carbon emissions.

Some visionary leaders at utilities that currently rely heavily on coal have taken the position that if we're going to impose taxes or caps on carbon emissions, it would be better to do so sooner rather than later. Instead of making expensive upgrades to aging coal plants to meet clean air regulations, only to find themselves having to reduce carbon emis-sions again later, they'd rather know about those carbon limits right now. That way they have a clear choice to replace rather than upgrade the plant.

We should take those utility executives at their word. Instead of continuing to fight costly political battles over how much and how fast the country's dirtiest coal plants should be improved, we should replace those plants with some combination of wind, gas, and nuclear power. And rather than indulge the fantasy that carbon emissions can be sharply cut by persuading literally hundreds of millions of drivers to radically change their ways, we should embrace the notion of replacing our 100 or 200 dirtiest coal-fired power plants with superior energy-generation technologies. That is not an inexpensive proposition, to be sure, but it's also not nearly as expensive as it may seem at first glance. Basically, it's like replacing an old clunker that you've driven for a very long time with a much better new car that you expect to also drive for a long time.

But to fully appreciate the force of this argument, we must first add up all the burdens of our present-day coal economy. The first part of this book (chapters 2 through 4) explores the social, economic, and global benefits and costs of burning coal.

PART ONE

COAL

*A Faustian Bargain
with Payments Coming Due*

(Overleaf) Western Kentucky Electric's coal plant, on Route 41, about 30 kilometers south of Henderson   *Source:* Tim Connor

The Basis of It All

Pennsylvania in the Pennsylvanian

TODAY, ON A RISE looming over Pittsburgh's city center, there is a historic marker proudly noting that the state's bituminous coal industry began on that spot on Mount Washington in 1760, and that the Pittsburgh coal bed was "eventually to be judged the most valuable individual mineral deposit in the United States." This is where the industrial revolution took off in the United States; it would run its course over two centuries, until the advent of our so-called postindustrial society. Now all the talk is of software services, technical innovation, and intellectual property. But what's sometimes lost sight of in that kind of chatter is that coal—yes, dirty, Victorian, pre-postmodern coal—still is the main fuel powering the United States. It generates more than half of the country's electricity—an even higher proportion than a generation ago—and remains key to many of the country's high-tech chemical and metallurgical enterprises.

It's true that most of the U.S. coal mining industry has moved west, to exploit rich, low-sulfur seams. But Pittsburgh, once more or less synonymous with Manchester as the essence of industrialization, is where U.S. coal mining began. And so Pittsburgh still is about as good a place as any to explore how coal got to be the basis of it all.

When the first European settlers arrived in what is now Pittsburgh, they recognized immediately that the area would be crucial to North America's future. Lying where the Allegheny and Monongahela rivers converge to form the mighty Ohio, the southwestern corner of Pennsylvania obviously would be the gateway to the Ohio Valley. First the French established Fort Duquesne at The Forks, and when the British took it away from them in the French and Indian (or Seven Years') War, in 1758, and built their own Fort Pitt, a Philadelphia newspaper said the acquisition laid open to the king's subjects "a vein of treasure which, if rightly managed, may prove richer than the mines of Mexico."[1] The vein referred to—figuratively—was the area's wealth of beavers.

Though some coal mining would begin just a few years after that, nobody could know as yet that The Forks were at the epicenter of a vast system of coal deposits, stretching east almost all the way to New Jersey, to the south and southwest into West Virginia and Kentucky, and to the west into Ohio, Indiana, and Illinois. Already by the end of the War of 1812, however, Pittsburgh was getting to be notorious for its dense smoke and gloomy skies. By the early 1830s, it was unique among American cities in running its many factories almost exclusively on coal; mills everywhere else still depended on hydropower. When the famous British geologist Charles Lyell visited the area in 1846, he was astonished "at beholding the richness of the seams of coal which appeared everywhere on the flanks of the hills and at the bottom of the valleys, and which are accessible in a degree I never witnessed elsewhere."[2]

At the time of Lyell's visit, there was still a lot to be learned about how Pittsburgh's rich coal seams had been created. By the previous century and probably long before, people mining and using coal had noticed that it often contained fossilized plants, and that fossilized remains of marine organisms—shellfish and corals, the likes of which were found living only in salt water—were often closely associated with coal deposits. But why would marine fossils be found in stratified deposits at the tops of mountains? What was one to make of fossilized tropical-looking plants found in temperate zones' coal, even in Pittsburgh's craggy peaks? If water somehow had played a role in all this, how could it have done so much monumental work in the mere six thousand years the Bible said the world had existed?[3]

From that first mining of coal in eighteenth-century Pittsburgh up to the 1960s, when the eastern coal industry and midwestern steel industries started to wind down, the Pittsburgh area would be at the heart of many monumental developments that made the United States what it is today: the industrial innovations that gave the North the manufacturing might to prevail in the Civil War; the great labor conflicts that followed soon after, pitting the nation's first proletarian leaders against the local magnates Andrew Carnegie and Henry Clay Frick; and finally, too, the birth of the twentieth century's progressive social contract, which can be reasonably dated to the settlement that President Theodore Roosevelt forced on the coal owners to end the first great coal strike of 1902.[4]

In that two-century period there was the stuff of legend, history, drama, and poetry. But how did the coal that made it all possible get to be there in the first place? There is drama too in the tale of how that story has come to be understood by the scientists who have devoted their lives to studying it.

If you drive the highways and byways in and around Pittsburgh, you can't help noticing a distinct layering in the various rock formations making up the walls of the many cuts that slice through the town's numerous hills. You only have to look a little more closely to see that certain rock types tend to reappear and alternate, and that there's even some regularity in the order of their appearance and disappearance.

In some other parts of the world, these repeating sequences are so regular they are compared to wedding cakes. In Pittsburgh, with three rivers converging, the topography is enormously complex, so that valleys interrupt what otherwise are neatly layered strata. The ups and downs bear comparison with the famous hills of Rome and San Francisco, observes John Harper, a Ph.D. geologist who has lived nearly all his life in Pittsburgh and now works for the Pennsylvania Geological Survey as chief of oil, gas, and subsurface services.[5] Harper was trained at the University of Pittsburgh by one of the foremost experts on western Pennsylvania's layered sedimentary rocks, Harold (Bud) Rollins. He now spends most of his time advising property owners and prospective buyers on whether there might be untapped oil and gas lurking under their garden statuary, but he welcomes the opportunity to explain to a visitor the extraordinary layered structures his teacher did so much to elucidate.

Driving down Pennsylvania Route 28, the Allegheny Valley Expressway, northeast of the city center, Harper stops his car for a close-up view of one road cut wall. At the bottom, he explains, there's a layer of sandstone topped with a rather amorphous reddish iron-rich rock known locally as the Pittsburgh red beds. It was formed from the area's ancient—very ancient—soil bed. Right above the red beds (see photograph) is a thin layer of blocky limestone called the Ames. It is found almost anywhere you cut into the mountains in western Pennsylvania, and because of the distinctive fossils with which it is richly endowed, it

A Pittsburgh Road Cut   Regular repeating strata of stone, called cyclothems, are somewhat apparent in this road cut along Pennsylvania 28, the Allegheny Expressway, outside
Pittsburgh. The lumpy layer protruding toward the bottom is the fossil-rich Ames formation, which marks the boundary between two geologic ages throughout southwestern
Pennsylvania. The dark layer two thirds of the way up is a mixture of coal and shale, a moist,
flaky stone formed from undersea mud.   *Source:* William Sweet

provides an excellent marker as you burrow down into the processes that
created these rocks hundreds of millions of years ago. Within the Ames
limestone is the boundary between two major rock sequences found in
western Pennsylvania, the Glenshaw and the Casselman formations, and
between two geological ages, known in North America as the Virgilian and Missourian.[6] Those ages make up the upper layer, or the most
recent part, of what's called the Pennsylvanian age in North America: it
began about 323 million years ago and ended about 294 million years
ago, and it is the latter half of what's known everywhere in the world
as the Carboniferous period—earth's great age of coal-making, as the
name suggests.

The Ames limestone, with the Glenshaw formation below and the
Casselman formation above, consists mostly of shells of tiny animals that
once lived in salty oceanic waters, among them corals and brachiopods,

the creatures superficially resembling clams that Harper studied for his doctoral dissertation under the supervision of Bud Rollins. Above the Ames in the Casselman, there's more rock formed from ancient soil (paleosoil)—these are the upper red beds—and above that is sandstone, then gray shale, then claystone, and finally, thin seams of coal and shale. Above the coal, more or less repeating the sequence below, there's more sandstone, then a mixture of sandstone, siltstone, and shale known locally as the Birmingham Member. In road cuts a little higher, there are more red beds above the Birmingham, and above them the other rocks characteristically found in the Casselman's wedding-cake sequences; farther down in deeper road cuts lie similar sequences associated with the Glenshaw.

When Harper pulls off the road, downhill in the valley just below an old abandoned strip mine, he picks away at some shale to reveal another thin seam of coal, mixed with greenish rock. The gray shale above and below, made from mud that settled in a marine environment and then was compressed by the new layers forming on top so that most of its water was extruded, is still—more than 300 million years after its formation—moist and flaky. The sandstone, with which it alternates here and elsewhere, is more like concrete. Harper picks some corals and snails from limestone found at ground level nearby; as one's eyes adjust, suddenly one sees the marine coral and snails everywhere in the rock, glinting like tiny jewels in the afternoon sun.

It's now pretty generally recognized that the area around Pittsburgh was located along the equator and had a tropical climate during the later part of the Carboniferous period. It lay on the banks of a vast but shallow inland sea, which stretched far to the west, from what is today Ohio, Illinois, and Kansas, and covered an area of about two million square kilometers. Compared to today, the North American continent lay on its side tilted to the east (see map, p. 16), and was part of a supercontinent geologists have dubbed Protopangea, so that it was connected directly to today's Great Britain and Eurasia. In effect, the vast coal beds of the American Northeast and Midwest that so impressed Lyell are of a piece with those found too in the British Midlands, Ukraine's Donets Basin, and northern China—the areas where successive industrial revolutions began.

The writer John McPhee has provided a vivid description of what Pennsylvania's coal-producing swamps would have looked like in the Pennsylvanian period:

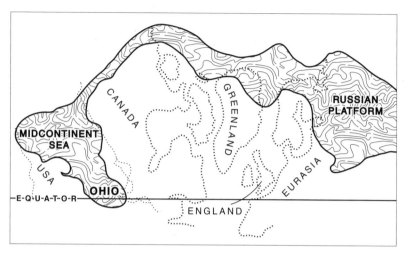

Ohio in the Pennsylvanian (approximately 325 to 290 million years ago)
Source: Heckel (2002)

Travelling west [toward Pittsburgh], and coming down from the mountains near Du Bois [in what is now the central part of Pennsylvania], you would have descended into a densely vegetal swamp [where big trees reigned]. They had thick boles and were about a hundred feet high. Other trees had bark like the bark of hemlocks and leaves like flat straps. Others had the fluted, swollen bases of cypress. In and out among the trunks flew dragonflies with the wingspans of great horned owls.... The understory was all but woven of rushlike woody plants and seed ferns. There were luxuriant tree ferns as much as fifty feet high.[7]

For the coal precursor peat to start developing in lush tropical marshes, there had to be a high water table and a steady infusion of fresh water, so as to support the furious vegetal growth needed to create a web that would prevent the rotting vegetation from being disrupted and dispersed by sediment carried in the waters that were fueling the plants. What is now western Pennsylvania met those conditions because water evaporating from the inland sea stretching up to Iowa and down to north Texas would form clouds, which in turn would produce abundant rainfall when they encountered the mountains to the east of the Pittsburgh area.

Pittsburgh in the Pennsylvanian often is described as looking a lot like today's Mississippi Delta, in which huge amounts of sediment are deposited by the river's forking branches to provide the raw materials for future rock. Sometimes it also is compared with the Gulf of Carpentaria, a semienclosed shallow sea on the northern rim of Australia just south of Papua New Guinea. Philip H. Heckel, a geologist at the University of Iowa, has estimated, based on both geological evidence and computer modeling of how Protopangea's climate would have been functioning in the Pennsylvanian, that North America's midcontinent sea had an average depth of about 90 meters.[8]

When scientists first trained their minds on the issue of precisely how Pennsylvania's coal-bearing wedding-cake layers developed, there seemed to be only two possible explanations for these puzzling repeating sequences of strata, dubbed "cyclothems" in 1932 by two of their first serious students, Harold R. Wanless and G. Marvin Weller of the Illinois Geological Survey. Either the coal-bearing cyclothems were created by delta shifting of the kind seen in the great floodplains of rivers like the Nile, the Ganges, and the Mississippi, so that marshlands were exposed at some times to the influence of seawaters and at others to the wearing down of hills and mountains; or the basins nurturing the marshes were somehow rising and falling periodically.

Looking back at scientific theories that have come to be discredited—like Ptolemaic astronomy and the phlogiston theory of combustion—it's often hard to make sense of them. It's as if the better knowledge that came later casts a shadow on what previously seemed clear, so that obvious objections stand out in sharp relief while reasons for believing the earlier theories are obscured. The early cyclothem theories are no exception. The deltaic model, which still has its adherents, seemed at first glance to account fairly adequately for rather irregular cyclothems of the kind often seen in Pittsburgh's road cuts. But even within the complex topography of the Pittsburgh area, the repetitive character of the cyclothems was a little hard to fathom. Why would peat marshes get covered up at regular intervals, sometimes with limestones formed in seawaters rising from the west, sometimes by sandstones made from the silt carried off the slopes to the east, so that the peat would be compacted into coal? Also, the time scales needed to account for cyclothems did not comport well with known deltaic processes: soils characteristic of cyclothems had matured too long, and been incised too deeply with valleys, for deltaic shifting to be the cause. In much of the midcontinental United States, where cyclothems were most extensive, topographic features had always been much more regular than in the Pittsburgh area, and evidence of deltas was simply lacking.

An alternative explanation for the cyclothems, involving subsidence and uplift of the earth's surface, was advanced by geologists like Weller. It was evident that over long periods of time, the earth's crust does indeed ripple. As newly formed mountains erode, silt is sent into basins, which then sink under the weight of the accumulating detritus, causing the aging mountains to rise again. The subsidence can be compared, observes Rollins, to the sinking of the earth's crust when ice accumulates, a well-known phenomenon observed in today's world. The periodic uplift of areas adjacent to basins might be explained in terms of less dense matter rising, or the curious but oft-noted fact that in a cylinder containing material of different shapes and sizes, the larger pieces tend to rise while the smaller ones fall.

Thus, the subsidence-uplift scenario, like the deltaic model, had a certain surface plausibility. But it also was more than a little murky in its details and was vulnerable to evidence that continued to mount throughout the twentieth century of the cyclothems' amazing geographic extent, regularity, and frequency.[9] They were discovered throughout land along the shores of the great inland sea: the present U.S. states of Ohio, Illinois, Iowa, Missouri, Kansas, Oklahoma, Texas, and New Mexico. When cyclothems were correlated in time over wide areas, by making use of special formations like the Ames limestone marking the boundary between Pennsylvania's Glenshaw and Casselman formations, they were found to recur as many as several dozen times. If account was taken for how the ground had shifted in the 300 million years since the Pennsylvanian, and sedimentary layers were traced cross-country, it emerged that a selfsame layer from the Missourian period might be referred to by the place name of Muncie Creek in Kansas, Millersville–La Salle in Illinois, or Portersville–Wood Run in Ohio and Pennsylvania.

What is more, as prospecting teams from companies like Esso (now Exxon) explored with new seismic techniques perfected in the 1940s and 1950s, hoping to find oil tucked into the unconformities often found in sedimentary sequences, they learned that the cyclothemic formations continued even undersea.[10] Corresponding formations, which could be cross-correlated to the North American and undersea cyclothems by means of fossil analysis, were identified in places as disparate as England, continental Europe, Russia, India, Australia, South Africa, and even China.

The most decisive argument against the subsidence-uplift account came, however, from time-frequency analysis. As scientists like Rollins and Heckel made ever more refined distinctions among types of cyclothems and came to ever stronger consensus estimates of the time the cyclothems would have taken to form, it became evident that subsidence and uplift could not have worked fast enough to produce the periodic sequences seen in places like the Pittsburgh road cuts. Subsidence uplift processes were well documented on the scale of 2-million- to 10-million-year cycles, but the characteristic cylothems in which coal is found had periods that hovered around 20,000, 40,000, 100,000, and 400,000 years.[11] The effect of this increasingly precise time analysis was to resuscitate and confirm a theory that seemed far-fetched, at the least, when it was first put forth in 1936.

That year, in the August 31 issue of the *Bulletin of the Geological Society of America*, Harold R. Wanless and Francis P. Shepard published a seminal article, "Sea Level and Climatic Changes Related to Late Paleozoic Cycles."[12] They postulated that the periodic coal and rock sequences through Middle America had been deposited as giant ice sheets advanced and retreated in the Southern Hemisphere during the Carboniferous period. As ice accumulated or melted, marine waters repeatedly fell or rose, leaving land sometimes covered in salt water, sometimes laid bare. This startling theory, dubbed the glacial-eustatic (the technical jargon for sea level changes induced by glaciation and deglaciation), depended in the first place on a detailed understanding of both physical and marine geology. Wanless (see photograph), the principal author, was a geologist at the University of Illinois at Urbana–Champaign who had done detailed cyclothemic mapping with Weller in the area around Peoria, Illinois. Shepard was a marine geologist associated with Woods Hole, the famous marine science laboratory on Cape Cod.

The glacial-eustatic model did not play too well initially among professional geologists, who found plenty to quibble about. For one thing, evidence of South Pole glaciation was rather tenuous. It was true that early in the twentieth century, the South African geologist A. L. Du Toit had detected the characteristic evidence of ice ages—tillites, moraines, and smoothly ground rock surfaces—in and around his sub-

Harold Wanless (*center foreground*) with students and colleagues in the field
Source: University of Illinois

tropical homeland.[13] As Wanless and Shepard put it in their article, Du Toit had shown "the late Paleozoic glaciation to have begun during the Middle Carboniferous, continued through the later Carboniferous and ended during the Early or Middle Permian."[14] The flora and fauna in the deposits, which Du Toit found in South Africa as well, were characteristic of cold climates. In due course, they would be one key to recognizing that Antarctica, South America, Africa, Madagascar, Australia, and India all were once joined in a supercontinent centered around the South Pole, which would come to be called Gondwanaland. Yet it was no mean trick to correlate such fossils with those found in the northern continents in similar periods: climatic conditions were radically different, so fossils found in each place were by and large completely different; and it was hard to get a good time stamp on the southern glaciers because, in a nutshell, later cataclysmic events in the south obliterated or blurred the boundaries by which their age could be determined.

Only with the publication of articles by the Texas geologist J. C. Crowell in 1978 and by the Australian geologists J. J. Veevers and C. McA.

Powell in 1987 were such technical problems satisfactorily resolved, so that the geologic mainstream became persuaded of a correspondence between southern ice ages and northern cyclothems.[15] A particularly crucial piece of evidence, says Rollins, was the discovery of a certain amoebalike organism that became extinct toward the end of the Paleozoic, not long after the Carboniferous, but was found in both the southern deposits and the cyclothems all the way from Pennsylvania to Oklahoma and Texas.

Just as important to the establishment of the glacial-eustatic theory, however, was the gradual acceptance of plate tectonics and continental drift;[16] as the new theory took hold in the 1950s and 1960s, Veevers and Powell, and like-minded researchers, used it to chart the extent that Gondwanaland glaciation might have achieved on minimal and maximal assumptions (see map). There now could be little doubt, however much deltaic processes or subsidence and uplift may have influenced individual features in individual places, that the dominant mechanism behind all the world's Carboniferous cyclothems was a succession of Southern Hemisphere ice ages. It was the only plausible explanation providing a common causative framework for the widespread patterns found in nature.[17]

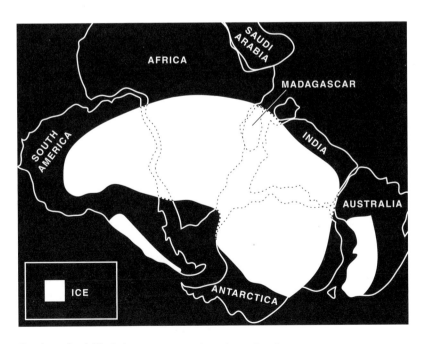

Gondwanaland Glaciation   *Source*: Crowley and Baum (1991)

It remained only for Phil Heckel to sum everything up, in a series of increasingly refined articles that he has been publishing since the late 1970s. In one of these, published in 1995[18] and specifically addressing the types of formations found around Pittsburgh, Heckel wrote that the marine deposits from the late Pennsylvanian resulted from deglaciation-driven incursions of the seas into the apron of the Appalachian highlands. "Because these marine units typically overlie coal beds, the main environment of formation of this type of coal bed appears to have been that of a coastal swamp migrating." Because the Appalachian basin was at that time quite distant from permanent oceanic sources of moisture, the moisture needed to form coal beds must have come from the midcontinent sea. Heckel estimated that this sea fell and rose about 90 meters, depending on whether water was getting locked up in southern ice or not.

Challenging critics to come up with a better theory or refurbish an old one, Heckel concluded in the same article that any challenge to the glacial-eustatic model would have to "provide an alternative abundant source of moisture to maintain large widespread peat swamps long enough to produce the thick coal beds in this landlocked basin, and yet account for the drier climates evident in paleosoils and nonmarine limestone that formed during lower sea-level stands."[19] Only the glacial model could "raise sea level fast enough to a depth great enough to account for the unique transgressive sequence of midcontinent Pennsylvania cyclothems…and to repeat these cycles with the frequencies required by the empirical data," he wrote in another article.[20] "All currently proposed tectonic models fail to change sea level fast enough, to a depth great enough, across an area broad enough and to repeat consistently within periods of time short enough to account for the cyclothems."

Though Heckel and Rollins, and other students of cyclothems, have had disagreements over the years about how best to classify them and what kind of terminology to use to characterize them, they are in full accord that only Gondwanaland glaciation can account for the coal-bearing cyclothems of the Pennsylvanian age. And they agree that the frequencies of those glaciation cycles strongly suggest that subtle changes in the Earth's physical relationship to the Sun must be the underlying causes of the Carboniferous ice ages—just as they caused the more recent ice ages, those of the last couple of million years. The cycles are named for the Serbian mathematician and engineer Milutin Milankovitch, who in the 1930s worked out the precise dynamics that explain their periods of 20,000, 41,000, 100,000, and 400,000 years.[21]

The cycles are associated with periodic changes in the tilt of Earth's axis, the shape of its orbit around the Sun, and the orbital timing of equinoxes and solstices. Basically, the Earth's long-term climate cycles as well as its short-term seasonalities are determined by three changing aspects of its relationship to the Sun: a rocking of its spin axis, so that its tilt (obliquity) relative to the Sun changes regularly; a compression and decompression of its orbit, so that at times it is more circular, at times more elliptical (or "eccentric"); and periodic changes in position in the Earth's orbit of the equinoxes and solstices and a periodic wobbling (as opposed to rocking) of Earth's tilt (elements of motion referred to as "precession"). Each of these aspects has a distinct cycle: precession, roughly 20,000 years; obliquity or tilt, 41,000 years; and eccentricity 100,000 and 400,000 years. These cycles show up in markers of Earth's ancient climate, its paleoclimate—not only in cyclothems but also in ice cores, ocean floor sediment borings, and so on.

The dominant element in determining seasonality and long-term cycles of seasonality is not the Earth's changing distance from the Sun—changes in eccentricity—but rather the wobbling of its tilt relative to the Sun—obliquity. The importance of that factor may not be intuitive, at least not to those of us who wrongly inferred, upon learning years ago about the elliptical orbits of the planets, that this is what makes the summers hot and the winters cold. Why would a slight change in how far Pittsburgh is from the Sun, as the Earth tilts one way or another, be more important than the much bigger changes associated with the shapes of Earth's orbit?

To get a feel for what's involved, think about the experience of tanning on a beach, where you get brown a lot faster if you are sitting up facing the afternoon sun than if you're lying flat, or contemplate the grapes growing on the slopes of Germany's Rhine and Mosel rivers, where those facing the sun produce sweet white wines, while those growing on flat land aren't good for much of anything. To understand this exactly, recall that light consists of particles called photons, and that any given area of Earth is irradiated by a precise number of them at any given time. If you were to shine a beam of light straight at a board, the circle illuminated would be bombarded by a certain number of photons; if you then tilted the board about 60 degrees, so that the circle formed an ellipse with an area twice as large, the photons would be spread over twice as large an area, so that any one part of it would get only half as much energy.

Put yet differently, the Sun always shines most directly at the equator, and as Earth circles the Sun, any given place above or below the equator sometimes tilts toward and sometimes away from the Sun. Above the Arctic Circle and below the Antarctic Circle, the Sun does not shine at all some days in midwinter. And over long periods demarcated by the Milankovitch elements, all these factors undergo subtle variations, so that irradiation is more or less intense in any given area—but especially at the highest latitudes, closest to the poles.

The upshot is that our current seasons are determined mainly by the Earth's tilt, with changes in orbit and precession (a periodic wobbling) of the solstices and equinoxes modifying the amount of radiation at any given place and time. Over long periods, sometimes the changes in tilt and distance work in tandem in one hemisphere or the other, so that seasonal changes are more or less dramatic in the south or north. At present, for example, changes in distance reinforce changes in tilt in the Southern Hemisphere but counteract them in the Northern Hemisphere. Thus, the differences in winter and summer conditions are more acute in the Southern Hemisphere and less acute in the Northern Hemisphere than when the opposite relationships prevail.

Heckel, who has done more than anybody else to precisely correlate the coal-bearing cyclothems of the Pennsylvanian age with the Milankovitch cycles, sees in the current alignment a cause for concern. With the North Pole's summers colder and drier than when the tilt and distance cycles are mutually reinforcing, there ought to be a big, steady buildup of Arctic ice.[22] Instead, the ice is melting, and melting fast. This is one reason to fear the earth may be headed for serious trouble on the largest of scales.

The world in which the vast Pennsylvanian seams of coal were created was an extraordinary place, in the most literal sense. The conditions that made it possible for that coal to come into being existed just once in earth's history. At the dawn of the Carboniferous, the first fernlike plants with spores had only just appeared, setting the stage for the riotous development of the webbed vegetation that would grow in the shade of giant *Lepidodendron* and *Sigillaria* trees. Throughout the tropical landmasses that later migrated into the Northern Hemisphere's temperate zones, dragonflies with wingspans of 30 inches buzzed about, while 6-

foot-long millipedes and 15-foot-long amphibians slithered along in the mud. Along the equator, whether the land was what is now the Donets, Westphalia, Manchester, or Pittsburgh, as the writer Simon Winchester has put it, "there were fetid and swampy jungles, all mud, dead ferns, and sagging branches of club mosses and horsetail."[23] Meanwhile, around the South Pole, what are now the huge landmasses of India, Australia, South America, Africa, Madagascar, and Antarctica had assembled into the supercontinent Gondwanaland, where a kind of supercontinental climate prevailed. Giant clouds from the surrounding oceans would migrate inland and, especially when the Milankovitch cycles were aligned so as to maximize moisture, send torrents of snow onto the continent's growing ice cap. As that water was sucked out of the oceans, sea levels would fall everywhere, probably by as much as 90 meters, leaving areas exposed throughout tropical Protopangea to form ancient soils and then—when the Milankovitch cycles were aligned the opposite way, and the waters returned—to host swamps and form peat that would then turn into coal under the layers of mud and seashells building up on top of it.

Of course, not all the world's coal was formed that way and at that time. Much of the world's bituminous coal was laid much later, in the Cretaceous period (140 to 65 million years ago), or even more recently. Soft coal (lignite) deposits in places like eastern Germany may have been put down just 10 to 12 million years ago. Sometimes, for example in eastern Pennsylvania, bituminous deposits from the Carboniferous era were later folded and compressed, as continents collided and mountains rose, to form anthracite—the hard coal that is almost pure carbon and free of contaminants. It burns more cleanly and gave cities that used it, like New York and Boston, better reputations for livability than the bituminous coal-using cities like Pittsburgh, Cleveland, Cincinnati, and Chicago,[24] and it made the United States uniquely well endowed for steel manufacturing.

In the relatively low-lying areas around the Pennsylvanian midcontinent sea, where seawater rich in iron sulfide combined with decaying plants, the coal would be high in sulfur.[25] Then, when the coal was burned hundreds of millions of years later, the marine-derived pyrite it contained would oxidize to form airborne sulfates and ultimately acid rain. Accordingly, as environmental regulations were imposed in the 1970s and 1980s to discourage use of high-sulfur coal, the U.S. mining industry moved west, to tap the huge seams of low-sulfur coal laid down in the Eocene, about 45 million years ago, without salty seawater playing a role. To the extent that the mining industry continued to work Appalachian fields, it went to the high ground, above levels reached by the Pennsylvanian midcontinent sea at its periodic high points; there, the miners lop off the tops of mountains to get at the low-sulfur deposits that formed under layers of nonmarine sediment, grind up the product, separate the coal, and fill dammed valleys with the sludge left over.

Wherever coal formed, it took hundreds of thousands or millions of years, at least, for peat swamps to be compressed to form seams of coal. However long it took and whenever and wherever it happened, the essential process in creating coal, always, was the extraction by plants of carbon from the carbon dioxide in the air. When that coal is then combusted—and in countries like England and Germany, virtually all the accessible coal has been burned in a period of less than three hundred years—the carbon is re-released into the atmosphere, where it acts like a blanket, retaining solar irradiation to make earth warmer. Even more than the transcendentalist Ralph Waldo Emerson appreciated, when he observed that coal "carries the heat of the tropics to Labrador and the polar circle," heat is indeed "a portable climate."[26] In effect, processes that took tens of millions of years to be completed are reversed in a wink of the eye, in geologic terms, with far-reaching consequences for our future on earth.

In the next century, conservative estimates suggest that the so-called greenhouse effect—the somewhat inaccurate shorthand generally used to refer to human-induced warming of the planet—is expected to cause an increase in temperature comparable to the temperature changes that separated the glacial and interglacial periods of the last two million years. Atmospheric concentrations of carbon dioxide are far higher today than they have been any time in the last million years, whether before, during, or between the ice ages. One of the main causes and arguably the one most readily corrected is coal combustion.

For reasons having to do with its basic chemistry, specifically the ratio of carbon atoms to hydrogen atoms, when coal is burned, about two to three times as much greenhouse gas is produced per unit of energy yielded as with oil or gas. With the ambiguous exception of burning wood,[27] which has an even more unfavorable ratio of carbon to hydrogen, no single thing is worse for the world's climate. From this point of view, combusting coal is a kind of excess—a pact with the devil—and payments are now coming due.

———

CHAPTER 3

The Air We Breathe

The Human Costs of Coal Combustion

THE REASONS coal has become the fuel most used to generate electricity in the United States—not to mention countries like China, and India, where it's even more dominant—are not hard to identify. Unlike oil, which must be imported from distant and untrustworthy foreign suppliers, it is available right here and readily recoverable in gigantic quantities. What is more, it will be in adequate supply for centuries to come. Most important of all, considered in a narrow monetary sense, burning coal is the cheapest way of generating electricity. As oil and natural gas prices skyrocketed, starting in 2003, coal's advantage has widened. This is why it accounts for well over half the electricity produced in the United States.

Coal's disadvantages, on the other hand, are largely hidden. The entire process of extracting coal and then disposing of waste products, which are hugely voluminous, is confined to just a few geographically and sparsely populated regions of the country. Under normal circumstances, only a tiny fraction of the U.S. population ever sees the coal industry in action. As for the emissions from coal-fired electricity plants, even though they are thought to cause thousands of deaths annually in the United States and hundreds of thousands of added hospital admissions, their effects are entirely statistical: old Uncle John may have a chronic respiratory condition that was fatally aggravated by constant exposure to coal emissions, but nobody ever says, "Poor old John. He died from breathing coal smoke."

The greenhouse gases associated with coal combustion—mainly carbon dioxide—are completely invisible. Their effects came to be generally recognized by the public only in recent decades, and even now, few people have any inkling just how drastic those effects are. The climate ramifications of coal combustion are the main theme of this book. But to think sensibly about all the advantages and disadvantages of coal, versus the alternative energy sources that will be considered in part 3, it's necessary first to have a complete view of coal's downside as well as its upside.

Anybody who has ever suffered a serious asthma attack, or watched almost helplessly as a child or aging parent struggled with one, knows the terror of not knowing for sure whether the next breath will be enough. Besides being enormously debilitating and requiring constant vigilance among chronic sufferers and those who care for them, asthma can and often does kill. When aggravated by particulates in the air, including aerosols formed from sulfur and nitrogen compounds, the condition is even more recurrent, debilitating, and frightening, and somewhat more deadly. The same is true of other medical conditions that can be compounded or even induced by exposure to severe pollution levels—upper and lower respiratory conditions of every kind, from minor colds to progressive bronchitis and fatal bouts of pneumonia, as well as cardiopulmonary conditions that can lead in the extreme case to cardiac arrest. On the hottest and most unpleasant summer days, when ozone alerts are declared throughout the eastern United States, the old and infirm are warned to stay inside and minimize activity. The ozone in the lower atmosphere that can stop their hearts is a by-product of power plant combustion and vehicle emissions. (This should not be confused with stratospheric ozone. Although chemically identical, it shields us from ultraviolet radiation and, until recently, was thinning dangerously as a result of reactions with chlorofluorocarbon gases used in refrigeration systems and aerosol spray cans.)

The kinds of noxious atmospheric conditions that can affect half the country at once fortunately are rare events. But in the most polluted parts of the country, where power plants are concentrated or traffic congestion is at its worst, dangerously high levels of pollution are not unusual, and the more astute physicians treating patients with conditions like asthma learn to watch out for them. Those places are not always where one might imagine.

Take Asheville, a pleasantly sleepy town in western North Carolina, on the edge of the scenic Smoky Mountains, best known to America at large as the birthplace of the novelist Thomas Wolfe. If one were to map the country's largest coal-burning utilities—Ohio's American Electric Power and Cinergy, Atlanta's Southern Company, and North Carolina's own Duke Power among them—and draw lines connecting all their coal-fired plants, the lines would all intersect in Asheville's vicinity. Accordingly, it ranks as one of the country's most chronically polluted cities. Wolfe famously said, "You can't go home again." But if you happen to be a less advantaged citizen of Asheville, and if you have the bad luck to suffer from asthma and find yourself showing up often in the middle of the night at a local clinic for emergency nebulizer treatments, you may wish you could just leave home and live somewhere else, anywhere else.

Cincinnati, Ohio, though not far at all from Asheville as the crow flies or the coal particulate blows, corresponds better to the average person's preconception of what a really polluted city is like. In the southeastern part of the state, on the Ohio River, Cincinnati indeed is one of the more difficult places to live and breathe freely. Dr. Jonathan A. Bernstein, an associate professor of clinical medicine at the University of Cincinnati's College of Medicine, reports that on high-smog days, he regularly sees more asthma visits and more patients generally suffering from shortness of breath, wheezing, and coughing.[1] On the very smoggiest days, he tells his patients to try not to come in at all because being outside may be more dangerous than going without treatment. And those very smoggy days are not uncommon, as emissions from local power plants tend to get trapped down low in the Ohio Valley during temperature inversions, in which cold air holding pollutants is trapped near the surface by a warmer layer immediately above, so that the normal process of upward convective diffusion is stopped. Though smog may be associated in people's minds mainly with traffic, and in places like Los Angeles or Washington, D.C. is in fact caused mainly by cars and trucks, in Cincinnati coal-fired power is overwhelmingly the source, says Bernstein. The pollutants blow in not just from plants in Ohio itself but also from those in neighboring West Virginia and Kentucky.

As chair of the American Academy of Allergy, Asthma and Immunology's committee on air pollution and author of a report about pollution for allergists, Bernstein knows what he's talking about. Yet even for a person of his experience, there is rarely, if ever, a case where the physician can say that pollution is the whole cause or even the main cause of a specific patient's condition. "If patients are allergic, there are interactions between particles and allergens, and it's very difficult to disentangle that, though you might consult the air quality index to get a sense of the situation," he explains. The morbidity and mortality connected with air pollution, like the diseases well known to be associated with tobacco use, are by nature statistical. Only by conducting large epidemiological studies, in which every variable that could be relevant is controlled to gauge the effect of the pollutant in question, can its impact be guessed.

One measure of coal's importance in human affairs is that the history of efforts to accurately estimate its baleful effects is virtually coterminous with the history of statistics itself. In 1662, the father of modern statistics, John Graunt, took it upon himself to closely examine London's health records as an exercise and demonstration of the nascent science.[2] The city had hundreds of thousands of inhabitants, and already had been fueled for centuries almost entirely by coal. As early as 1285, King Edward I had established a commission to evaluate air pollution, and twenty-two years later he tried to ban coal burning in the city altogether.[3] Centuries later, a London visitor or dweller observed that "by reason...of the Smoak," the "Air of the City, especially in the Winter time, is rendered very unwholesome: for in case there be no Wind, and especially in Frosty Weather, the City is covered with a thick Brovillard or Cloud, which the force of the Winter Son is not able to scatter... when yet to them who are but a Mile out of Town, the Air is sharp, clear and healthy."[4] So Graunt was not the first to suspect that the smoke could not be good for people and other living things, and sure enough, when he tallied up his numbers, he estimated that one fifth to one fourth of all the deaths in London each year were the result of lung-related diseases.

Two hundred years later, when the abolitionist sisters Catherine and Harriett Beecher Stowe published a book about home economics, they worried about how domestic happiness was affected by the three or four tons of coal they guessed it took to heat the average American family home for a winter.[5] Yet the string of causes and consequences connecting coal smoke with deadly or debilitating ailments like lung cancer, asthma, emphysema, and heart failure still seemed speculative and abstract. It took the terrible London smog of December 8, 1952, and four years before that, a more limited but just as dramatic health emergency in Donora, Pennsylvania, to start driving home the lethal impact of air pollution. Both events were induced by temperature inversions. In London, particulate and organic compounds combined with particulate sulfur dioxide and nitrogen oxides from domestic and industrial coal combustion, to form a noxious brew. The number of deaths above the number that normally would have been expected in London during the week right after the inversion was estimated at 2,800.[6]

In the case of Donora, the trapped pollutants came almost entirely from the batteries of coke ovens that were the lifeblood of this company town, located just southwest of Pittsburgh. The immediate death toll on and right after October 26, 1948, the day Donora suddenly was smothered by almost unbreathable air, was eighteen.[7] The longer-term impact remains uncertain to this day, partly because Pennsylvania's head of public hygiene dismissed it as a "one-time atmospheric freak,"[8] partly because statisticians were still developing the refined techniques needed to distinguish deaths that occurred prematurely or need not have occurred at all from those that would have happened as a matter of course.

As late as the early 1970s, when a professor at Pittsburgh's Carnegie Mellon University made one of the first comprehensive attempts to estimate the exact health effects of coal combustion, he still found the science politics rough sledding. Although Lester Lave was (and is) one of the country's foremost experts on the U.S. electric power industry, and although he submitted his work to a top mathematician at Princeton University, John Tukey, for a close critical review of his statistical methods, he nonetheless ran into considerable hostility, both from industry and from other epidemiologists. Just the same, the pathbreaking article he coauthored with Eugene Seskin and published in *Science* in 1970 was perhaps the first to definitively prove a causal relationship between pollution and death and disease.[9]

Lave and Seskin first surveyed previous work on the subject, then presented results of their own "cross-sectional" survey comparing U.S. cities with varying levels of pollution. They found that a 50 percent cut in urban pollution would reduce mortality and morbidity from bronchitis by 25 to 50 percent, all respiratory disease by 25 percent, and the cost of all cancer care by 15 percent. They estimated that the combined savings in health-care costs associated with the 50 percent reduction in pollution would be about $2 billion in 1970 dollars. They noted, however, that this was a tremendous underestimate of real total costs, because it didn't account for earnings lost as a result of death and disease, let alone the value individuals placed on their own lives.

The Lave-Seskin study came under attack on technical grounds because it didn't correct for personal habits like smoking and because of possible false correlations: critics pointed out, for example, that big cities tend to be more polluted than smaller ones, and that people allegedly tend to die at higher rates in the bigger cities because of a supposed "urban effect." Critics also complained that the study contained no "longitudinal" analysis—that is, evaluation of how mortality and morbidity changed over time with varying pollution levels. Lave and Seskin did include longitudinal analysis and controls for personal habits in a comprehensive book they published in 1977, *Air Pollution and Human Health*,[10] but, by Lave's account, they were getting tired of the whole controversy, and critics often didn't seem to notice that their complaints had now been addressed.

The work by Lave and Seskin established an ironclad link between pollution and health. Yet to this day, observes Devra Davis, an eminent but controversial epidemiologist who happened to have grown up in Donora, "there has never been a surgeon general's report on air pollution"[11]—something comparable, that is, to the 1964 report that established a connection between smoking and ill health, and that led to the long but ultimately quite successful campaign to discourage tobacco use.

Nevertheless, two studies conducted in the early 1990s pretty well did what a surgeon general's report might have accomplished, not only providing persuasive evidence of the links between pollution and morbidity and mortality, but also yielding precise estimates of their magnitude. One, conducted by researchers with Harvard University's School of Public Health, focused on six cities with varying levels of air pollution. The other, sponsored by the American Cancer Society, was national in scope.[12] These studies prompted the Environmental Protection Agency to promulgate an emissions reduction standard for fine particulate (defined as 2.5 parts per million), which, the EPA estimated, might save about a quarter of the 60,000 American lives lost each year as a result of exposure.

Particulate pollution comes from numerous sources, including industrial processes that rely on coal and diesel vehicles. But coal-fired power plants top the list. The Clean Air Task Force, a Boston-based nonprofit advocacy group, determined in an October 2000 report that power plants outstripped all other polluters as the main emitters of sulfur dioxide, which is the biggest single source of fine particulate pollution in the United States, and were the major source of nitrogen oxides, the other main fine-particulate precursor. The task force claimed, on the basis of calculations done by independent consultants and closely based on the EPA's own models, that if power plants were required to reduce their emissions of sulfur dioxide and nitrogen oxides by 75 percent, the effect would be to save about 18,000 lives a year—roughly the same number lost annually in drunk-driving accidents. That conclusion implied that the total number of deaths attributable each year to the two pollutants is about 30,000.[13]

Starting in 1970, with the creation of the Environmental Protection Agency by President Richard M. Nixon and the drafting of the country's landmark Clean Air Act by Nixon administration officials, the United States has made a concerted effort to clean the air. It has made much more progress in this endeavor than most other advanced industrial countries, which have tended to emphasize energy conservation more than mitigation of air pollution. But when President George W. Bush took office in January 2001, the struggle to bring coal-burning electric power plants into stricter compliance with clean air regulations was at a critical juncture.

During the 1970s and 1980s, largely as the result of a "cap-and-trade" system set up to reduce the acid rain that was sterilizing the country's rivers and lakes, output of sulfur dioxide—the main precursor to acid rain—had been cut by more than a third (see "Cap-and-Trade Versus Pollution Tax," below). That dramatic reduction was achieved at much lower costs than industry and independent analysts had expected. By allowing utilities for which reductions were too costly to purchase emissions credits from those that found compliance easier, the system introduced a flexibility appreciated by all parties to the clean air debate. The U.S. Environmental Protection Agency estimated that sulfur dioxide output was 40 percent lower in 1990 than it would have been without the cap-and-trade system.[14] During the next five years, from 1990 to 1995, emissions from the 261 most polluting power plants in the country—those required to come into compliance with 1990 clean air amendments first—dropped by another 45 percent, from 9.7 to 5.3 million metric tons.[15] That was achieved mainly by switching to lower-sulfur coal: in 1990 it accounted for about two thirds of coal utility

CHANGES IN ANNUAL U.S. COAL PRODUCTION FROM 1990 TO 1995

North Appalachia high-sulfur	−29 million tons
Illinois high-sulfur	−40 million tons
Wyoming/Montana low-sulfur	+78 million tons
Colorado/Utah low-sulfur	+10 million tons
Central Appalachian low-sulfur	+15 million tons

Source: U.S. Energy Information Administration

generation; just five years later, it was well over three quarters, as western displaced eastern coal (see table).

Impressive progress also was made in bringing the two other pollutants from power plants under control. Emissions of particulate were cut in half during the 1970s, while nitrogen oxides stayed flat, with reductions compensating roughly for increases that otherwise would have occurred. But progress slowed in the late 1990s, as utilities took advantage of a loophole in clean air legislation. They were required to see to it that new plants and equipment complied with target emissions levels, but if utilities could make a plausible claim that they were making routine improvements in existing facilities, then they could avoid installing costly scrubbing devices. The whole effort to continue cutting sulfur and nitrogen emissions came to a standstill over the issue of "new source review." This is the bureaucratic name for the regulatory process in which utilities and regulators sparred over whether plant upgrades were routine and therefore exempt from clean air requirements, or major.

Meanwhile, mercury loomed as an additional pollutant crying out for regulation, and environmentalists and industry were locking horns over whether greenhouse gas emissions also should be called pollutants and regulated as such. The two main greenhouse gases, carbon dioxide and methane, certainly were not noxious in the normal sense—at the concentrations being put into the atmosphere, they were not damaging to anybody's health, or even, unlike acid rain, to ecosystems or physical structures. But their growing presence in the atmosphere was well known to be gradually warming the planet, at a rate that was truly alarming. From 1990 to 2002, U.S. output of greenhouse gases increased 13 percent; the power sector accounted for a disproportionate share, with its emissions rising at least 25 percent.[16] (Though more up-to-date data is unavailable at this writing, the rate of increase in power plant emissions has certainly been even greater since 2002, as sharply rising oil and gas prices have prompted U.S. utilities to burn more coal than ever before.[17]) If greenhouse gas emissions continue to grow that fast, it will be tough by any reckoning for the United States to live up to the pledge made at an international conference held in Kyoto in 1997 to get its carbon dioxide emissions well below 1990 levels by 2010 (see "What Is the Kyoto Protocol?" below).

Cap-and-Trade Versus Pollution Tax

In a cap-and-trade system, a ceiling is set for the total amount of a pollutant that can be emitted in a country or region in a given period of time. The economy is divided by industry or sector, and companies or organizations known to be releasing the pollutant are issued permits allowing them emissions up to some proportion of the total ceiling. All participating organizations may trade the allowances freely, so that those finding it easier to stay below their maximum level can sell allowances to those finding it harder. The system reduces the aggregate cost of meeting emissions targets and introduces a flexibility that industry likes.

This works relatively well for pollutants that are quickly dispersed in the atmosphere over long distances, so that it does not make much difference locally where emissions originate. Cap-and-trade was first implemented in the United States to reduce the chemicals that turn into acid rain, and it worked nicely. The system works less satisfactorily for heavy chemicals like mercury that tend to be deposited near their source: if your own food is contaminated with mercury from nearby sources, it obviously will not help you out if the emitter buys a permit from some other emitter to release even more into the environment. Cap-and-trade systems are very effective in controlling the main greenhouse gas, carbon dioxide, which mixes rapidly in the atmosphere, so that everybody is helped by emissions reductions made by anybody anywhere. The European Union has adopted a cap-and-trade system to meet the targets its member states have accepted under the Kyoto Protocol.

Arguably, however, a carbon tax is an even more efficient way of reducing carbon dioxide emissions. In keeping with free market principles, such a tax establishes a completely level playing field among all economic sectors and penalizes those organizations in exact proportion to the carbon they emit, without much need of central planning and sectoral partition. Because of coal's greater carbon output per unit of energy, compared with oil, such a tax affects the coal industry about twice as sharply as the automotive sector.

The Kyoto Protocol was adopted as a first step toward implementing the International Framework on Climate Change, a treaty formulated at Rio de Janeiro in 1992 that the senior George Bush had signed on behalf of the United States. But the protocol was closely associated with the controversial views and persona of Vice President Al Gore, and in signing the country on to it, President Bill Clinton plainly got too far ahead of public opinion. The Senate made clear in 1998 that it had no intention of ratifying the agreement, and in the 2000 presidential campaign, George W. Bush explicitly repudiated it.

The grounds for opposition to the protocol are easily identified. Bringing the United States into compliance with Kyoto would put an even bigger burden on the power sector than is evident at first glance. According to a 2001 governmental report prepared as part of the Kyoto negotiation process, coal-fired plants operated by U.S. utilities were responsible for 29 percent of carbon dioxide emissions (compared to the transportation sector's 26 percent). And that didn't include the growing production of electricity by generators other than utilities. Factoring them in, the report said, coal-fired power production accounted for close to 40 percent of U.S. greenhouse gas emissions.[18]

Small wonder, then, that in the 2000 election campaign, Vice President Al Gore—despite having played a big role in putting global warming on the political agenda and in midwifing the Kyoto Protocol as well—scarcely mentioned the subject. After all, he represented the high-sulfur coal state of Tennessee (which he managed to lose anyway), and it was taken for granted that other coal states like Ohio and Illinois would be decisive in the election outcome. In 2004, challenger John Kerry, despite a strong record on environmental issues, produced an exact repeat of Gore's performance. Again, it was taken for granted that coal-burning Ohio would be decisive, as indeed it was. Bush, for his part, did not conceal his disdain for Kyoto, which he said would disadvantage the United States in global trade; after winning the presidency in 2000, he lost no time rejecting it. He indicated in the run-up to the election that as president he would treat carbon dioxide as a pollutant to be regulated along with sulfur dioxide, nitrogen oxides, and ozone—but after taking office he repudiated that commitment too.

The Bush administration, in fact, watered down every kind of clean air measure during its first term. Notably, it drastically weakened plans to sharply curtail mercury emissions, and after a lengthy bureaucratic

What Is the Kyoto Protocol?

The basic principles of the Kyoto Protocol were adopted at a conference in November 1997, but it took three more years of international negotiations to hammer out the agreement's complex details. The nations of the world were divided into so-called Annex 1 countries, which were required to meet certain greenhouse gas emissions targets by 2008–12, relative to a 1990 baseline, and developing countries, which are not subject to any emissions requirements in that period.

Confusingly, descriptions of the protocol often give different numbers for the total emissions reductions that industrial nations are required to make by 2010. This is probably because the targets differ for each country, and in some cases—recognizing many special circumstances—increases in emissions actually are allowed. Norway, Iceland, and Australia are permitted to increase emissions by as much as 10 percent, for example, and no cuts are required of Russia, New Zealand, or the Ukraine. The United States, had it ratified the agreement, would have been required to reduce its emissions by 7 percent. The European Union agreed to reduce its combined emissions by 8 percent.

The protocol covers six greenhouse gases: carbon dioxide, methane, nitrous oxide, hydrofluorocarbons, perfluorocarbons, and sulfur hexafluoride. The degree to which each of these gases absorbs infrared radiation (their "global warming potential") varies extremely widely, offering opportunities to achieve targets by tailoring reduction strategies to particular gases. Methane has a greenhouse impact that is 23 times that of carbon dioxide, per molecule, and the impact of the hydrofluorocarbons is as much as 12,000 times that of carbon dioxide.

The protocol and its implementation agreements enable industrial countries to meet their targets partly by taking measures that absorb greenhouse gases, for example by planting new forests. The protocol recommends emissions trading among countries, an approach developed and favored by the United States, to meet global targets. Industrial countries also get credit for projects they foster in developing countries to reduce gases. The developing countries are required to start inventorying their greenhouse gas emissions, but they were exempted from the protocol's first-phase targets in recognition of the fact that rapidly growing societies cannot realistically promise to make cuts or even predict what targets are achievable.

Critics in the United States have complained that developing countries are getting a free ride. However, by not participating in the Kyoto system, the United States is benefiting free of charge from greenhouse gas emissions cuts made—often at considerable inconvenience—by the countries that have accepted it.

battle that ended up costing Bush's first EPA administrator her job, the government announced its intention to ditch the new-source review process for sulfur dioxide reduction. First the administration tried to get the Department of Justice to drop enforcement suits against seven top midwestern and southern utilities; when Justice refused to do so, the EPA simply walked away from settlements that the utilities were about to accept. In August 2003, the Bush administration raised the threshold at which new source rules would kick in so much that the remaining senior air pollution enforcement officials from the Clinton administration took early retirement in disgust. A *New York Times Magazine* writer concluded, "The administration's real problem with the new-source review program wasn't that it didn't work. The problem was that it was about to work too well."[19]

Enforcement of regulations for mercury abatement also was weakened, according to a *Wall Street Journal* report, following a November 2003 meeting in which representatives of coal-burning utilities talked administration officials into relaxing the rules. A Clinton administration proposal would have required mercury emissions to be cut by 90 percent by 2007–8; the Bush plan called for 70 percent cuts by 2018.[20] About a third of the mercury that gets into the air each year comes from coal-fired power plants, which emit about 50 tons annually, making them the biggest single source.[21] Another third is from municipal and medical wastes, but those sources are being sharply curtailed. Regulation of mercury emissions from coal plants, on the other hand, remains hotly contested, to put it mildly.

Given the endless haggling that the new-source review process had engendered, perhaps a good argument was to be made for a different approach. Utilities in the most highly polluting states of the Midwest and Southeast had been the targets of endless litigation brought by the northeastern states most affected by sulfur dioxide, nitrogen oxides, and ozone blowing in from across their borders, without much tangible result. Bush proposed to replace the new-source process with a cap-and-trade system like the one that had proved successful in cutting acid rain. This seemed a good idea, except that the target dates set for compliance in his proposed "Clear Skies" program were set far into the next decade, much later than the targets originally established in the Clean Air Act and amendments that had set the new-source process in motion. Bush proposed no provision for tightening caps over time.

The debate over clean air regulation took a surprising turn in March 2005, when a Republican-controlled Congress narrowly rejected Bush's Clear Skies program. Almost immediately, the EPA issued a new plan for sulfur and nitrogen abatement, the Clean Air Interstate Rule, which represented a compromise between the original Clinton program and Clear Skies. Sulfur dioxide would be cut 70 percent from 2003 levels by 2015 and nitrogen oxides 60 percent, at a total estimated cost of at least $36 billion. The EPA predicted the program would ultimately result in 17,000 fewer premature deaths annually from air pollution, and 1.7 million fewer days lost to pollution-related illnesses. The new interstate rule was widely hailed as a step in the right direction. But a similar compromise plan for mercury abatement, issued days later, met with much sharper criticism from environmentalists and public health specialists. It too established a cap-and-trade system, which is a suitable method for reducing pollutants like sulfur dioxide or carbon dioxide that mix well in the atmosphere and are widely and evenly dispersed, but arguably not for reducing a heavy element like mercury that tends to stay concentrated in "hot spots." A deal made to trade mercury abatement in one area for continued pollution in another in effect leaves all the pollution in that one area.

Generally, while public health has taken a back seat in the Bush years, the administration has never dared say it is against cleaning the air on principle; its *modus operandi* has been to proceed in stealth mode, claiming that it is stretching deadlines to make compliance more realistic and to make the whole process more efficient. Reducing greenhouse gas emissions, however, is another matter. The president has said repeatedly that this is simply something the United States cannot afford.

The inventory of adverse coal impacts does not stop, of course, with the air pollutants and greenhouse gases. Mine safety and environmental issues associated with coal extraction also are major issues. Though mining fatalities are no longer a great scourge, as when more than a million Americans dug deep into the ground, the shift to strip mining and mountain lopping has brought new horrors: vast stretches of wondrous landscapes in Montana and Wyoming turned into desolate moonscapes on earth; beloved local scenery changed beyond recognition in West Virginia, Kentucky, and Tennessee, where hilltops are removed, crushed, and filtered, with the waste sludge dammed up in valleys, creating a whole new risk to public safety (see photographs).

Strip-mining operations on Kayford mountain, a half hour east of Charleston, West Virginia. Seams of coal can be seen in the exposed stone face, with a coal company truck barely visible (*top, right*). Across the valley, below the cemetery where generations of coal miners are buried (*middle, left*), lies a flat plateau, ringed with some trees and covered with grass (*bottom, right*), which used to be a mountain looming over the cemetery. The man in the middle photo is Julian Martin, a retired teacher and mining activist.   *Source:* William Sweet

Altogether, there are about 700 coal slurry dams or "impound-ments" in the United States, about 200 of which are built over aban-doned mines. On February 26, 1972, one such dam gave way in Logan County, West Virginia. Some 132 million gallons of sludge suddenly flooded the Buffalo Creek Valley floor, destroying 17 communities, killing 125 people, and leaving 4,000 homeless. In terms of volume, though thankfully there was no loss of life, the biggest such disaster oc-curred on October 11, 2000, in Inez County, Kentucky. Yet consider-able evidence suggests that upon taking office several months later, the Bush administration's approach to the problem of the impoundments was to demote engineers in the Mine Safety and Health Administration who were best qualified to address it, to stack the agency's leadership with former mining executives, and to generally obstruct investigations into the Inez County disaster to protect responsible parties from civil and criminal liability.[22]

The administration has taken an essentially similar approach to the problem of mountaintop lopping. In May 2002, Judge Charles H. Haden II of the Southern District of West Virginia declared in a rul-ing that the administration's revisions to rules governing the practice represented an obvious perversion of the Clean Water Act. "The rule change was designed simply for the benefit of the mining industry and its employees," he said.[23] Yet in January 2004, the administration pro-posed further relaxation of the rules: lifting a Reagan administration prohibition on simply dumping mountain detritus into stream beds. The Interior Department's proposed new rule would allow the practice, provided water quality was protected "to the extent practicable."

As a matter of course, coal combustion generates millions of tons of waste—fly ash, bottom ash, and boiler slag—that must be disposed of each year. Ironically, when emissions are scrubbed for sulfur by means of flue-gas desulfurization (FGD) or electrostatic precipitation, the result is millions of tons of extra solid waste to be got rid of, and increased generating expenses. (FGD adds as much as 8 percent to electricity prices.[24]) Thus, scrubbing coal emissions trades a public health hazard for an admittedly lesser environmental blight, at signifi-cant monetary cost.

Although all that is serious enough, coal's most worrisome envi-ronmental effect is on the global atmosphere. Even though coal-fired power accounts for only about a quarter of U.S. energy consumption while oil—almost all of it used to fuel automotive vehicles—accounts for 40 percent, their contribution to total greenhouse gas emissions is roughly the same. This is because of oil's greater chemical efficiency in terms of carbon emitted per unit of energy produced, and the relatively low average generating efficiency of the nation's aging power plants.

The correspondences between the oil-fired transportation and coal-fired power sectors are slightly uncanny, a bit like the similarity in size of the Sun and Moon as seen from Earth—that is to say, essentially coincidental, but helpful as memory aids and analytical devices. As noted, estimated yearly deaths from power plant emissions are at least as great as total yearly deaths from drunk-driving accidents, and possibly as great as total traffic fatalities. Each sector, though quite different in terms of how energy is converted and used, contributes about a third of the nation's total greenhouse gas emissions. And each sector, in principle, could make equal contributions to reducing those emissions, which are putting the future of the planet at risk. But doing something serious about automobile emissions runs up against America's love affair with the car, and particularly the gas-guzzling SUV. And doing something serious about coal runs up against the immense political power of the midwestern and southeastern coal-burning utilities, an obstacle traditionally believed to be just as high as or even higher than the automotive lobby—though that view may be mistaken.

The coal industry is not in fact the immense political force it was fifty years ago, when the United Mine Workers of America numbered more than a million members and its militant leaders, like John L. Lewis and Phil Murray, could threaten to shut down the U.S. economy if their demands were not met. Today, the UMWA has barely more than 100,000 members, many of them retired.[25] Scarcely 70,000 workers actually mine coal, and a great deal of that is stripped by means of huge machines, operated by a handful of nonunion, highly skilled and highly paid men and women. From this perspective, coal might seem almost a spent political force, the stuff of nostalgic songs sung by aging folk-song performers.

From a different perspective, however, the coal industry still exercises almost the same disproportionate sway over the U.S. polity as in the UMWA's glory days. This is because the big utilities in the Midwest and the Southeast rely on it utterly. They are the nation's (and among the world's) very largest utilities, with names like American Electric Power (the mightiest of them all), Southern Company, Duke Power, and First Energy. AEP, First Energy, and Cinergy all are located in Ohio, where, increasingly, the closely divided nation sees its destiny decided every four years.

Some measure of the utilities' influence and power can be taken from the events that led to the great Midwest–Northeast electricity blackout of August 14, 2003. That event darkened states from Michigan to New York, as well as Canada's Ontario, making it the largest single outage in history. Though the failure was rooted in the deregulation and restructuring of the U.S. power system, which began in earnest in the early 1990s, both the underlying and the proximate causes could be traced mainly to the negligence of one Ohio utility, First Energy.

The immediate chain of events leading to the August 2003 blackout began two years earlier, when the Davis Besse nuclear power plant operated by First Energy, near Toledo, had to be closed down for detailed inspection and reconstruction when unexpectedly severe corrosion was discovered in the vulnerable cap to the reactor core, which is pierced with control rods and fuel rods. Because the situation was so serious—if the corrosion went too far, the reactor's pressurized vessel might burst, releasing vast quantities of radiation into the environment—the Nuclear Regulatory Commission had to order emergency inspection of sixty-nine similar reactors, at considerable expense and inconvenience.[26] Those reactors were found in due course to be all right, but meanwhile, First Energy's Davis Besse plant stayed shuttered, resulting in a shortage of electricity right in the middle of the narrow corridor that connects the midwestern and northeastern power systems.

Beginning early in the afternoon of August 14, big transmission lines began to fail in First Energy's operating area, several because the utility had not kept up with tree-trimming, so that as heavily loaded lines heated up, they sagged into brush and shorted out. As one went down, the next would become too loaded, sag still more, and short, and so on. All that, the result of a serious infringement of operating standards and no small matter in its own right, would have remained a local problem if First Energy and the midwestern power regulator had quickly recognized what was going on and had promptly cut service to enough customers to keep the whole system from getting overloaded. But the equipment First Energy needed to monitor and simulate what was going on in its system was out of order, and the situation at a newly established regulator in Indiana was not much better. Six months later, when a U.S.–Canada investigatory team reported on the accident, a list of the ways in which First Energy was seriously unprepared for the events that unfolded on August 14 filled the better part of a page. A second list, of the ways in which the utility had violated standard reliability rules, filled another page.[27] Yet there was no talk of imposing civil or criminal penalties. As if nothing noteworthy had happened, the midwestern and southeastern utilities continued to successfully resist federal legislation that would have made reliability rules mandatory and strengthened the hand of the Federal Energy Regulatory Commission. In particular, they forced FERC to back off from imposing a "standard market design" requiring all U.S. utilities to play by the same set of rules.

And so, if you ask yourself why burning coal continues to kill tens of thousands of Americans each year, why it still causes neurological disorders in hundreds or thousands of children, why it continues to ravage environments from the Smokies to the Tetons, and why it produces 40 percent of U.S. greenhouse gas emissions each year and 10 percent of the whole world's emissions—and if you ask why so little is done about all that—you need look no further than the amazing events of August 2003 and the role Ohio famously plays every four years in each presidential election.

In recent years, however, the alliance of midwestern and southeastern coal-burning utilities has shown signs of fracturing on global warming. In essence the situation is similar to that in the global oil industry, where companies like BP (British Petroleum) and Shell have broken ranks with the mainstream, taking the position that the energy industry is going to have to find ways of weaning the world from carbon-based fuels. Already in the late 1990s, Ohio's AEP, probably the country's largest utility at that time, began to cautiously favor carbon regulation. More recently, James E. Rogers, CEO of Cinergy—the Cincinnati-based utility that emerged as an industry giant in 2005 after merging with Duke Power—has adopted an aggressive public position similar to that taken by BP's Sir John Brown. Both Rogers and Duke CEO Paul Anderson have been saying that global warming is a real and very serious problem, and that energy companies can survive in the long term only by addressing it. With Rogers, Brown, and Anderson, personal conviction and vision are clearly playing an important part, but ultimately they are acting in what they see as their corporations' self-interest.

Assessment of such interests is a complicated matter. One factor, already mentioned, is that as the United States had adopted strict regulations limiting emissions of sulfur dioxide, utilities burning western low-sulfur coal have had an advantage over the eastern utilities relying heavily on Appalachian coal. But if carbon emissions are regulated as well, the western advantage is partly canceled, because eastern coal burns more efficiently and therefore emits less carbon per unit of electricity generated.

A larger consideration is that as utilities make expensive upgrades to aging coal plants to meet clean air regulations, they worry that if carbon is to be regulated as well, it might make more sense to just replace the plants rather than improve them. Since electricity generated by natural gas is cheaper than coal-generated electricity under most circumstances, and electricity generated by nuclear power plants or wind farms is only marginally more expensive, studies have indicated that if carbon emissions were taxed, a very large fraction of the U.S. coal industry would be promptly shut down.[28]

Last but not least, large institutional investors tend to buy stock in utilities and have an exceptionally large influence on their management. In recent years, many of those institutional investors have been showing up at annual shareholder meetings and demanding that the managers of coal-dependent utilities prepare formal plans for somehow transitioning away from carbon. For example, when shareholders gathered in Tulsa, Oklahoma, in April 2005 for AEP's annual meeting, an activist group warned that the country's number-one carbon emitter risked relinquishing leadership on carbon to Cinergy and Duke. The preceding year, in July 2004, the attorneys general for eight states, including New York, California, Iowa, and Wisconsin, filed suit against a group of energy organizations for producing 10 percent of U.S. carbon emissions: Cinergy, Southern Company, Xcel Energy, AEP, and the Tennessee Valley Authority.

The coal industry's future is hanging in the balance. It should be determined by conscious decisions taken in the broadest public interest.

CHAPTER 4

From Outer Space

Asia's Brown Cloud, and More

Veteran astronaut Richard Truly, speaking a number of years ago to fellows of the National Academy of Engineering, said that on repeated visits to outer space over the years, with each look back at earth, "the effects of human activity are dramatically apparent."[1] Truly, who went on to be the head of the U.S. government's National Renewable Energy Laboratory in Golden, Colorado, was referring to a sight that observant airline passengers will have noticed too. If you fly, say, from Chicago to Boston, from Frankfurt to Stockholm, or from Tokyo to Beijing, all along the way you'll see smoke rising from tall stacks, most of which are venting emissions from coal-fired power plants. Seen from higher up, from a U.S. space shuttle or Russia's Mir space station, those columns of smoke converge into larger plumes, and then sometimes into gigantic blankets of smoke and dust that cover whole regions. The most massive and portentous of those, which, oddly, was identified barely more than a decade ago, is a brown cloud that sometimes blankets the entire Asia-Pacific region, from India to China and Japan and down to South Pacific islands like Fiji and the Maldives.

When Truly said the impact of human activity is disturbingly evident from space, he was referring in part to the visible material known to be an acute health hazard as well as a threat to ecosystems, agriculture, and physical structures. That material, consisting of sulfur dioxide and nitrogen oxides, black carbon, white ash, and a toxic brew of miscellaneous other aerosols, emanates in massive quantities from the U.S. industrial heartland, northern Germany, and southern Scandinavia; the sprawling metropolis around Saõ Paulo, Brazil; and many similar megacities shooting up all over northeastern Asia. Much of it also is made up of combustion products from biomass—sticks, straw, dung, and other agricultural by-products—burned in the developing countries of Asia and Africa. Lesser but still significant constituents come from natural sources, notably dust and volatile organic compounds breathed out by some species of trees.

But what Truly also had in mind, most particularly, was that the visible matter converging into massive clouds is a sign of something even more insidious: the greenhouse gases known to be warming the earth alarmingly. The greenhouse effect, as such, is natural and benign, and now is familiar to all educated people: gases like water vapor and carbon dioxide in the atmosphere prevent some of the earth's heat from radiating back out into space, making earth warm enough for mammalian life to evolve and be sustained. But additional greenhouse gases that have been pumped into the atmosphere since the industrial revolution began have been enhancing that natural effect, perhaps to a dangerous degree. Carbon dioxide, from both combustion and animal respiration, is the best known and most important of the gases associated with human activity. But a variety of other gases—nitrogen oxides; methane, which comes from sources like rice paddies, animal excrement, and natural gas distribution lines; and chlorofluorocarbons (CFCs), which once were used ubiquitously in refrigerators and pressurized cans—also are quite significant.

Because of the discovery that CFCs destroy the stratospheric ozone layer that shields earth and protects life from ultraviolet radiation, they are being phased out in keeping with the Montreal Protocol of 1987, one of the most effective international agreements adopted to protect the global environment. Though the work of diplomats and the result of lobbying and pressure from a multitude of concerned citizens all over the world, the protocol can be attributed ultimately to the atmospheric chemists who recognized the subtle and devious ways in which CFCs, as well as nitrogen oxides, consume ozone. It was a singular instance of new knowledge getting translated promptly into good policy, and the scientists who produced that knowledge—Mario Molina, Sherwood Rowland, and Paul Crutzen—were fittingly honored with the Nobel Prize.

By the time they got that recognition, another scientist, an extraordinary Indian engineer turned atmospheric chemist by the name of Veerabhadran "Ram" Ramanathan, had discovered that CFCs, besides wrecking the ozone, also are one of the most potent greenhouse gases.

He showed that per molecule, they produce about 10,000 times as much warming as carbon dioxide. Accordingly, the phase-out of CFCs under the Montreal agreement has produced a big fringe benefit for the climate—a little noted fact with some big implications that will bear revisiting as the various dimensions of the climate problem are further explored.

Ram Ramanathan grew up in a small village near Madras (now officially known as Chennai), in southern India, a member of the usual extended family in which grandparents or great-grandparents presided over a sprawling clan of brothers, sisters, cousins, and unmarried aunts. His grandmother, whose home provided a summer retreat for Ram and his immediate family, cooked—like tens of millions of other Indian housewives—mainly with cow dung. Ramanathan remembers that soot from the burned dung blackened their kitchen walls.[2]

A strong pupil, Ramanathan earned an engineering degree from one of the region's polytechnic institutes and went to work for a local company, where he was given the job of developing containers, ironically, for CFCs. He hated working in the corporate world and soon made his way to the United States, to study at the State University of New York at Stony Brook, Long Island. There, he earned a Ph.D. in 1974 in the study of planetary atmospheres, seemingly a big step from CFC cans. Immediately his career took off, and soon he would be producing scientific work of the greatest importance. As with so many of his brilliant compatriots who attended American universities, became successful, and then sought ways to be of some use to their mother country, he would find his work relevant to the lives of those he had left behind.

In 1975, just a year after getting his doctorate, Ramanathan showed for the first time that CFCs are a potent greenhouse gas and a key factor in the earth's total greenhouse budget—the balance of gases and aerosols (suspended particles) that either trap heat radiating from the earth, making the globe warmer, or reflect solar radiation, making it cooler. (Ramanathan says now that even he was astounded to discover that the CFCs, even though they are but a minuscule fraction of the gases in the atmosphere, are so potent that they have an appreciable effect on the total radiation budget.[3]) Three years later, he made a second major contribution to the field, showing that the stratospheric ozone being consumed by the CFCs also was a significant greenhouse gas.

In 1979 Ramanathan helped design, and then became principal scientist—basically the manager—of the Earth Radiation Budget Experiment, in which the objective was to use data from earth satellites to measure and assess the earth's inflows and outflows of radiation. Over fourteen years and with a cumulative budget of $150 million, ERBE would collect a wealth of data from three NASA satellites. By the end of the 1980s, the project showed that the earth's cloud cover has a net cooling effect on climate: the amount of solar energy that clouds reflect back out to space far exceeds the amount of energy they trap. At the same time, ERBE data measured the extent to which water vapor—gaseous water in the atmosphere, as opposed to the water droplets that make up clouds—warms the planet, exerting an undesired feedback effect. The warmer earth gets, the more water evaporates from the oceans; the more vapor there is in the atmosphere; the warmer the oceans get; and so on.

Over its twenty-five-year life span, ERBE has taken the greenhouse effect from the realm of theory and computer modeling to workaday empirical science, showing that it can be measured directly and accurately. So Ramanathan, in just the first fifteen years of his scientific career, firmly established two previously unknown important facts about the greenhouse budget, directly measured the budget itself, and identified what stands out as the biggest single quandary in all of climate science: the net role of clouds. As the earth gets hotter with increases in greenhouse gases, will increased cloud cover dampen the effect or make it even more potent? Ramanathan calls this "the Gordian knot" of the climate problem. At the end of the 1980s, he wrote an overview of cloud science for *Physics Today* magazine,[4] a monthly read by almost all physicists in the United States and many overseas as well, in which he delineated the main elements of the conundrum. If he were to write that article today, he says, it would be hardly different.[5] None of the decisive issues has been resolved.

In the next decade, however, it's likely that many of the uncertainties associated with clouds and climate will be reduced or even eliminated. A set of earth-observing satellites known collectively as the A train, after the famous New York City subway line immortalized in a Duke Ellington tune, is being launched in coordinated fashion, all to follow similar orbits and cross the equator at roughly 1:30 p.m. mean local time.[6] Aqua, Aura, CALIPSO, and CloudSat already are in orbit, systematically gathering observations of the earth's water and energy cycles, atmospheric chemistry, aerosols, and clouds. Aqua alone carries a half dozen sophisticated instruments that produce data being analyzed by scientists all over the world: for example, the Moderate Resolution Imaging Spectroradiometer (Modis), the Humidity Sensor for Brazil (HSB), and Clouds and the Earth's Radiant Energy System (Ceres). The Franco–U.S. satellite, CALIPSO, contains just two major instruments, one to measure the way aerosols polarize light, the other to do infrared imaging of atmospheric processes. The initial four satellites will be joined by Parasol, which will further evaluate the earth's radiative budget, quantify the atmosphere's reactions to anthropogenic gases, and track its evolution in time. An Orbiting Carbon Observatory, to be launched in 2006 or 2007, will monitor the carbon cycle.

By the second decade of this century it ought to be possible, given all the data that will be delivered by the A train and a number of other planned satellites, to make much more exact statements about how clouds of different kinds at various heights affect climate, and how human activities, clouds, and climate interact in real time. Not only that, the instruments will elucidate certain large-scale regional phenomena that are well known to be affecting the whole world's climate, but in complicated and poorly understood ways. That includes deforestation fires and changed land-use patterns in Brazil's Amazon. And at least as important, the satellites will cast light on the huge brown cloud that covers India, Bangladesh, Southeast Asia, China's industrial heartland, and much of the Indian Ocean, mainly between the months of December and April—another major Ramanathan discovery, and the one most closely linked with his name.

The usual story goes that Ramanathan first noticed the big brown cloud over Asia and the Pacific on flights between India and the United States, but the reality is of course a little more complicated. Why, after all, didn't anybody else notice something so big and brown? Why didn't other world travelers, not to speak of astronauts and cosmonauts, see it? The short answer is that Ramanathan noticed it because he was primed to look for it. In the course of the ERBE study, his attention had been drawn by a kind of supergreenhouse effect.[7] When Pacific Ocean temperatures went above 27 degrees Celsius, suddenly the rate at which the temperature rose would accelerate, as water began to evaporate faster and faster. But at almost exactly 29.5 degrees, the temperature would abruptly stabilize. Why? Ramanathan and one of his students theorized that perhaps a kind of natural thermostat was at work. They thought that as the oceans heated up, there might be much more production of cumulonimbus clouds—the kind that Americans and Europeans think of as thunderheads—and that these might reflect sunlight back into space, so that the ocean surface cooled.

Ramanathan designed an experiment in 1993, working from a base in Fiji and using data from aircraft, ships, and satellites, to test that theory. The results were startling. The amount of sunlight reaching the surface of the ocean was much smaller than expected, which set Ramanathan's mind working on a different track. On flights from India and the United States to the Maldives and Fiji, he had noticed a brownish haze over the Pacific, and it occurred to him that the haze might be blocking sunlight from reaching the sea's surface. He supposes that pilots must have noticed the haze too, but they probably attached no special significance to it. As for astronauts, because the surface of the brown cloud is light and reflective, it's not especially obvious from outer space. You have to be actually in it to see it clearly.

Together with Crutzen, the Nobelist who had discovered the deleterious effects of nitrogen oxide emissions, Ramanathan named the haze the "Asian Brown Cloud," or ABC, and the two men set about establishing a major new long-term experimental program to study it. Crutzen recalls that one afternoon at Scripps, looking out over the Pacific Ocean, they realized that it would be worth studying the haze of pollution wafting into the Indian Ocean from the Indian subcontinent (which, in hindsight, he considers slightly hilarious, inasmuch as they were having the conversation under another brown cloud, spreading from Southern California westward into the Pacific).[8] Crutzen and Ramanathan dubbed their new enterprise the Indian Ocean Experiment (INDOEX), and in the next decade, it firmly established the cloud's huge scope and outlined its main origins and causes. The United Nations Environment Program (UNEP) asked Ramanathan to do a summation report, which led to the launching of the even broader Asian Brown Cloud (ABC) program, with UNEP funding.

In the course of INDOEX, the brown cloud was found to consist largely of gaseous pollutants like carbon monoxide, which came mainly from biomass burning, and aerosols, mainly from coal combustion. And it was found to be the combined product of certain characteristic weather patterns making for large-area inversions and of home and commercial heating during winter—whether that involved burning straw and sticks in the traditional grass-roofed huts of Ramanathan's ancestral home, or burning coal, oil, or gas in the skyscrapers being erected by multinational corporations in cities like Bangalore and Shanghai.

The cloud, two miles thick at its winter maximum, absorbs solar radiation before it can reach the surface and reflects it back out to space, so that on average, about 10 percent less sunlight reaches the surface of the earth than otherwise would be the case. Because of that effect, the cloud locally cancels the warming impact of the greenhouse gases arising from the same biomass and fossil-fuel combustion that generates the cloud. In a narrow sense, that might be considered a net good. But the cloud also has been found to have a highly deleterious effect on rainfall patterns, essentially because it cools the waters below, reducing rates of evaporation and cloud formation.[9] In their most recent published work, Ramanathan and his colleagues have done simulations of the brown cloud's effects from 1930 to 2000, in which they were able to replicate changes in surface solar radiation, surface and atmospheric temperatures over land and sea, and decreases in monsoon rainfall. Their work suggests that if current emissions trends continue unabated, the frequency of droughts on the subcontinent may double in the coming decades.[10]

Independent studies show that the brown cloud has been aggravating or even causing a drought that has been ravaging northwestern China for years, drastically reducing flows into the Yellow River, while shifting rainfall to the south. Similar droughts, but with a more patchwork pattern, laid waste to large swaths of India in 2003 and 2004. Ironically, by this time, the Indian government had dissuaded UNEP from providing extra funding to allow the ABC program to increase its scope to cover all of Asia. Evidently the country's officialdom, according to a long front-page report in the *Wall Street Journal*, did not wish to see aspersions cast on economic activities it considered vital to its progress.[11] In a way, this didn't matter. Ramanathan, by then a tenured full professor of ocean, atmospheric, and climate sciences at the Scripps Institution of Oceanography at the University of California–San Diego, was too prominent an authority to silence or even constrain. Besides being a member or fellow of just about every important scientific society, from the U.S. National Academy to the Pontifical Academy, he was well ensconced as Scripps as director of its Center for Atmospheric Sciences.

Something as big and pervasive as the Asian brown cloud cannot be attributed to just one cause or blamed on just one or two bad actors. Nevertheless, at first approximation, it consists mostly of products of coal and biomass combustion in China and India. Though China's controversial big dams, like Three Gorges on the Yangtze and Xiaolangdi on the Yellow River, get a lot of attention outside China, more than four fifths of the country's electricity is produced from coal. Coal also is the preferred fuel for home heating and cooking in much of the country. This is especially true in rural areas. Public authorities have sought to encourage conversion to natural gas in the bigger cities, but even there, coal is still very widely used. The hexagonal briquettes, perforated vertically with holes so as to better mix oxygen and carbon, are sold ubiquitously by vendors on China's street corners (see photo).

The energy picture in India is essentially similar, except that its economy is only about half as big and its extremes are more pronounced. Nuclear power plays a somewhat greater role, partly because the central government promoted it starting in the 1950s, for reasons connected with international prestige and a desire to lay the foundation for a weapons program. Yet India's rural population still relies almost entirely on biomass for fuel, and by and large it is coal that runs India's factories, powers its electricity generators, and drives its quaint locomotives.

Environmentalists in the United States and Europe may fret about whether air quality is getting better or worse, or whether it is getting better fast enough.[12] In China and India, the situation is unambiguous. As these countries strive to attain higher standards of living, their hunger for energy gets fiercer all the time—though their per capita consumption is still a tiny fraction of that in advanced industrial countries. And as their energy requirements escalate, so too does their coal combustion. Though they may want in principle to equip their coal generators with state-of-the-art pollution-control technology, it's almost a foregone conclusion that these two relatively poor but very ambitious countries tend to economize on such equipment. Even when they have properly outfitted plants, they often do not use it when demand for electricity is highest and there is pressure to run plants at maximum capacity.

The result is a public health catastrophe. A careful and reputable scholar at Pittsburgh's Carnegie Mellon University, Keith Florig, has estimated that as many as a million people die each year in China from diseases that are closely related to air pollution.[13] If that figure is correct, about twice as many Chinese die yearly from air pollution as die

Charcoal briquettes are used everywhere as domestic fuel in
China. *Source:* William Sweet

in accidents of all kinds. The World Bank's estimates are lower—on the
order of several hundred thousand premature deaths each year, attributed to air pollution mainly from coal and biomass combustion, and
from cement production.[14] In any case, it's generally recognized by all
who care to look that China's air pollution is not just a problem but a
public health emergency.

Vaclav Smil, an expert on energy and the environment at the University of Manitoba, has observed in one of his many China studies that
exposure levels in the country's towns "call to mind conditions in the
cities of Western Europe and North America two to four generations

ago."[15] Bjørn Lomborg, the Danish specialist on global trends who has styled himself "the skeptical environmentalist," does not gloss over the fact that Beijing, like Delhi and Mexico City, often suffers from particulate levels that are eight times those in Europe or the United States. He notes that of the fifteen most polluted cities in the world, thirteen are in Asia.[16]

In China, indoor air pollution from coal and biomass combustion often reaches levels deemed unsafe for *outdoor* air pollution by authorities like the World Health Organization. So if you live in or near one of China's most heavily polluted industrial cities, you may not do yourself any good, in contrast to your counterpart in North Carolina or Ohio, by staying inside when it's too nasty outside. An especially tragic result of this indoor air pollution, which is often associated with inefficient and poorly vented cooking stoves, is that the country's women—and evidently this is uniquely true of China—die of lung cancer at rates about as high as those for men, even though they smoke much less.[17]

In India, the picture is just as grim. According to the world's leading authority on indoor air pollution, Kirk R. Smith of the University of California–Berkeley's School of Public Health, between 400,000 and 550,000 women and children under five die prematurely each year as a result of fumes from domestic biomass combustion.[18] Smith considers that number a pretty firm low-end estimate of the death toll from indoor air pollution because it leaves out men, many of whom smoke—counting them too would produce much greater uncertainties in the calculations, but clearly the total mortality would be much higher.

According to Smith's studies, about three quarters of Indian families rely on poorly designed, unvented stoves, which produce pollution far in excess of world standards. In the few areas like West Bengal where coal is used instead, the toxins spewed into homes may be even worse. Thus, altogether, some 750 million rural Indians are systematically poisoning themselves, and that leaves out the whole modernizing part of the economy, the urban and industrial sectors relying on coal to produce electricity and steel. When this is factored in, India's toll from air pollution must be similar to China's—perhaps as high as a million people a year, and certainly hundreds of thousands. (Ramanathan guesses from everything he has read that the air pollution toll for each country is in the hundreds of thousands a year, and for Asia as a whole at least 1.5 million.[19])

Almost all of those premature deaths can be traced either to coal or to biomass, for which coal is the generally preferred substitute. At the same time, emissions from coal and biomass are the major constituents of the Asian brown cloud, which is having adverse climatic effects, at least locally. And because combustion of coal and biomass produces more carbon dioxide per unit of energy than any alternative fuels, continued reliance on them also contributes significantly to global warming. To be sure, if biomass is replenished as sticks and straw are burned, the net effect is carbon-neutral: the growing crops absorb the carbon dioxide emitted when the old crops are combusted. But as the Chinese and Indian economies modernize, which inevitably involves some permanent clearing of land without replenishment of crops or grass, biomass combustion also comes to have a net negative impact on regional and global climates.[20]

To its credit, China's government is not oblivious to the country's air pollution crisis. Besides prodding municipalities to switch to cleaner and more efficient fuels, notably natural gas and oil, Beijing has systematically encouraged hundreds of millions of farmers to buy more efficient cooking stoves. It also has sought to phase out the highly polluting small boilers that were the hallmark of the "town and village enterprises" that sprouted like mushrooms with the introduction of free markets under Deng Xiaoping, Mao's reformist successor. The State Environmental Protection Administration in Beijing set limits for the major pollutants like sulfur dioxide, nitrogen oxides, and particulate, and in the late 1990s, its media-savvy head would often show up at a plant site with camera teams in tow to denounce a "yellow dragon"— his term for a tall stack spewing out toxins.[21] The country even boasts a nascent environmental movement, which is small but independent and sometimes surprisingly aggressive.

Inevitably, though, in a decentralized and sprawling country like China, much depends on the honesty and zeal of local officials, many of whom are reluctant to restrict industries they see as their area's lifeblood and, not infrequently, are simply on the take. The corruption of China's state and local officialdom is legendary. An environmental bureaucrat representing the central government in Beijing recalled inspection tours he had made in Guangdong Province, the booming region around Guangzhou (formerly Canton) and near Hong Kong, in which the problem got worse the lower one went in the pecking order. "The more local the [environmental] board, the richer they got," he remembered, taking it for granted that the lower officials were absconding with fees levied against environmental scofflaws—money that was supposed to be plowed back into environmental improvements. "The provincial one had a dingy office and was poorly equipped. The board from Guangdong city was somewhat better. But when I went to the Ping Yu county board, they picked me up in a luxury sedan and took me to a brand-new air-conditioned office building!"[22]

It's of course not just local officeholders who want those air conditioners. Go to Shanghai and look out on the famous Bund. One of the most prominent neon displays will be for Carrier, the top U.S. supplier. And air conditioners are not the only modern convenience fervently desired. In the streets of Shanghai and in every other major city of the industrial heartland, automobiles already have largely replaced the bicycles that blanketed all the roads just twenty years ago. Everywhere, in all facets of life, equipment and devices that depend on fossil fuels are replacing the old. It could hardly be otherwise in a country proud of its achievements, eager to improve its material condition still more, and determined to reclaim a preeminent place in the world.

To judge from what the government's environmental officials say, privately as well as publicly, China would like to be not just a rich and powerful member of the world community but a good global citizen too. The Beijing government was the first to complete a survey of its greenhouse gas emissions, as required by the Kyoto Protocol, and it recognizes that the activities and processes producing those emissions also are fouling its cities, killing its people, and affecting regional ecology. But in trying to reduce the twin burdens of pollution and greenhouse gases, China faces terrible quandaries, as does the other developing giant, India.

Even if these two huge countries were growing much more slowly—in recent years they've been growing at rates of close to 10 percent yearly—they do not have the luxury of simply deciding not to burn coal anymore. Holland, in the 1960s, decided to wean itself from coal when newly discovered natural gas reserves proved a much more attractive option. The United Kingdom started to give up coal as a matter of policy in the 1970s, with the discovery of North Sea oil. (Today it is beginning to deploy offshore wind farms on a grandiose scale, following the example of Denmark and drawing on technology developed for high-seas oil extraction. Basically Britain is opting for large-scale wind energy because it prefers that to a new round of nuclear reactor construction, even though the latter course might be cheaper.) The United States, taking cues from them, also could and should give up coal and adopt a variety of low-carbon or zero-carbon technologies.

But the technologies and fuels that rich countries like the United States can easily afford are often too expensive for fast-growing poor countries like China and India. In the last five years, for example, China has invited foreign nuclear reactor suppliers to bid for construction of several very large power plant complexes. But nuclear energy, which produces no air pollutants to speak of and no greenhouse gases, comes at a distinct cost premium—the exact figure is very hard to pin down, but nuclear electricity is at least 10 to 20 percent more expensive than average electricity. So while it may be an attractive substitute for coal in fully industrialized countries, where electricity sectors are growing slowly if at all, it is not going to be much help in a country like China, which is adding the equivalent of England's entire power sector every two years.[23] China and India can barely afford to build the additional power plants they need just to keep growing at the impressive rates they want; they clearly want to keep the costs of providing that added capacity to a minimum.

What goes for nuclear energy also goes for hydroelectricity, which is, if anything, even more controversial among environmentalists—not only globally but also in India and China. The construction of China's Three Gorges Dam on the upper Yangtze River divided the country's political elites like no other issue, even before Mao died and the reform era began; in India, the noted novelist Arundhati Roy has spearheaded passionate opposition to construction of any new big dams.[24] Like nuclear energy, dams provide, for all practical purposes, zero-carbon electricity, but there are premiums to be paid both monetarily and environmentally.

From an environmental point of view, wind turbines and solar energy have the potential to be much more than just niche players in China and India. Especially in remote areas, where there is no electricity grid or centrally generated electricity has to be delivered over very long distances, wind and photovoltaic cells are appealing alternatives. Saving on transmission of electricity compensates for the higher installation costs of wind and solar facilities.

Even in the most optimistic projections for renewable energy technologies, however, it's clear that wind and solar cannot come anywhere close to supplying the additional electric power that China and India will need in the next decades, let alone make a dent in the coal combustion that's already killing millions and warming the world. But what about substituting lower-carbon oil and natural gas for coal? Clearly, it would be desirable for these countries to switch as much as possible from coal to oil and gas, which burn more cleanly and generate much less greenhouse gas per unit of energy produced. But as China and India furiously negotiate long-term development and supply contracts for oil and gas with nations in Central Asia and the Middle East, the effect is to drive up global prices, making it all the harder for them to afford replacing coal with less carbon-intensive fuels.[25]

Faced with the extreme difficulty of bringing down their carbon emissions as their economies rapidly advance, India's and China's policy makers might be tempted to decide that it's really not necessary after all, inasmuch as the regional warming effects of carbon dioxide are roughly balanced by the cooling effects of the Asian brown cloud. But that doesn't really work, as health policy or climate policy. Cutting the pollutants that are killing millions each year and making life seriously unpleasant for hundreds of millions more is an inescapable moral imperative. It's not just a matter of air pollution, either. In the advanced industrial countries, coal mining is by and large no longer a dark and dangerous business that takes hundreds of thousands of lives, and blights hundreds of thousands more, each year. But in China and India, it's still dismal work done in places "dark as a dungeon," as the American country song puts it, amid conditions reminiscent of those described by European writers like Emile Zola and Friedrich Engels at the end of the nineteenth century. Thousands die mining in each country every year.[26]

The World Bank has estimated that if all the added health costs associated with coal combustion in China were taken into account in setting the price of coal, it would cost twice as much as it does. Adopting policies to actually drive up coal prices to reflect social costs would powerfully encourage adoption of more efficient technologies and conversion to alternative fuels. But the World Bank also has cautioned that a significant decline in China's coal use in the next couple of decades, given current technology, "can only be achieved at enormous financial cost."[27]

In the United States, and to a lesser degree in countries like Australia and Canada, opposition to the Kyoto Protocol has centered largely around a feeling that its requirements are unfair because China and India—unlike the advanced industrial countries—are not required to cut emissions to meet definite targets. Realistically, however, there's no chance that these two fast-industrializing countries can cut their total emissions anytime soon. "It's a problem of decades," observes Ramanathan—meaning it will be at least twenty years before China or India will be in a position to actually stop the increase of greenhouse gas emissions, and then reverse it. This is not because they are self-indulgent or indifferent to the social and global costs of coal combustion. On the contrary, they are quite acutely aware of the costs; and as caste-ridden India becomes more socially conscious and authoritarian China becomes more democratic, that awareness will only sharpen.

If global warming were merely a matter of the world's getting incrementally a little warmer, with the advantages and disadvantages more or less evenly distributed, then it might make sense to carp about the fairness of each country's contributions to slowing the process. But the effects, though impossible to predict with complete precision, will be very uneven. And it's not a foregone conclusion that they will be merely incremental. Climate scientists have found in the last fifty years that climate change can be sudden and cataclysmic. Changes in greenhouse gas levels associated with past climate catastrophes have been smaller than the changes we are inducing now. We don't and can't know whether what we are doing to the atmosphere today could bring on a climate cataclysm in the lives of our children or grandchildren, but we cannot dismiss the possibility.

PART TWO

CLIMATE

The Lockstep Relationship Between
Carbon Dioxide and Temperature

(Overleaf)
Above: Original photograph taken in 1928 of the Upsala Glacier in Argentina
Below: The same scene today
Copyright © *Greenpeace / Beltra / De Agostini*

CHAPTER 5

The Drillers

DRIVING ALONG the south side of Cape Cod, on Route 28 from South Yarmouth to Chatham, where the cape abruptly curves north, you might notice, coming out of Harwich, a little antiques shop set back from the road, nestled between a country barber shop and a bait-and-tackle store. If you went in, you wouldn't find much at first glance to distinguish it or its proprietor, Chester Langway, from many other such establishments in the vicinity. There's a miscellany of the usual oddities—pots and pans, an old dentist's chair minus the drilling apparatus, glass and ceramics, and a depiction of Daniel Webster, New England's political leader in the first half of the nineteenth century, in which Langway takes special pride. If you get into a chat with Langway, a vigorous man in his mid-seventies who gets around in a little Ford pickup, he'll explain in his telltale New England accent—the wide vowels, the errant or missing *r*s at the end of words—how Webster used to frequent a raw bar in Boston, still a going concern, where he'd order four or five dozen oysters along with a fifth of bourbon to wash them down. He'll tell you he's thinking of donating the picture of Webster to the bar. Yet even if you linger a while, it isn't likely you'll find any clues to what Langway himself did before retiring to the cape, unless, at the very back of the store, you happen upon some soapstone carvings made by Inuits, whom Langway, indifferent to the "politically correct" conventions of the day, still calls Eskimos. Though the little figures are surely the most original items in the store, Langway evidently isn't too eager to sell them. They're the one memento of his unique contribution, in an unusual career, to what one historian has aptly called "an extraordinary revolution in paleoclimatology."[1]

Fifty years ago, Langway's life took a singular turn. After serving several years in the military, he had just earned bachelor's and master's degrees in geology at Boston University when, in 1956, he was hired by one Henri Bader to work as a research scientist at the U.S. Army Corps of Engineers Snow, Ice, and Permafrost Research Establishment (SIPRE) in Wilmette, Illinois, a suburb just north of Chicago. Bader, considered by Langway and others to be the father of ice core research in the United States,[2] was a native of Switzerland who had obtained his doctorate at the prestigious ETH (the Confessional Technical High School) in Zurich, the country's equivalent of MIT, and had worked at Switzerland's top ice laboratory in Davos, the Alpine town now best known as the spot where the industrial world's elites meet for annual confabs and as the fictional setting of Thomas Mann's *The Magic Mountain*. Bader, like Mann, sensed in the 1930s that things were going badly awry in Europe and began to cast about for some other continent to work on. His wanderings took him first to Moscow, then to various places in Latin America, where he found employment intermittently as a manager of mining operations. Finally, after World War II, he landed a professorship at Rutgers University in New Jersey, and then, still more improbably, the position of chief scientist with the U.S. Army's ice program. When he recruited Langway, he was getting ready to do some exploratory drillings into the ice sheets of Greenland and Antarctica.

Those sheets, covering more than 10 percent of the earth's land surface and containing nearly four fifths of its fresh water, represented—together with the ocean's depths and outer space—the planet's last physical frontier from a scientific point of view. In 1930, the great German meteorologist and geophysicist Alfred Wegener, the discoverer of continental drift, had died on a meteorological expedition in central Greenland. On that journey, one of his assistants, Ernst Sorge, noticed, upon digging a pit in the ice, that annual layers of accumulation could be discerned, as if they were tree rings. That discovery opened enticing possibilities. While there had been attempts, as Langway would later observe, to drill mechanically into Swiss glaciers as early as the 1840s to do thickness measurements, and later into the ice sheets, "never until now," in the mid-1950s, "had the snow-ice mantle of high polar ice sheets been completely penetrated."[3] There had been exploratory drilling in Alaska and Greenland in 1949, 1950, and 1951, in which Bader was involved, but the equipment kept breaking and jamming, and nobody got farther than about 100 meters into the ice. Nobody had ever attempted to read the world's paleoclimate by drilling long cores and examining the layers Sorge had discovered.

At the time Bader hired Langway, an international program of polar research was being organized, which would come to be known as the International Geophysical Year of 1957–58. Bader clearly recognized the potential in Sorge's discovery. "Two thirds of the area of the Greenland ice sheet and practically all of the Antarctic ice sheet are permanently dry," he wrote in a report on the polar ice and snow to be studied. "All precipitation is in the form of snow. Summer melt is rare, and usually affects a surface layer only a few centimeters thick [but marking the boundaries between annual layers]. Thus every snowfall, including everything that fell with it, is, so to say, separately and safely filed for future reference by being buried under later snowfalls."[4] To obtain those files, one had to drill down using a rock drill adapted to remove ice chips by means of a compressed air system or, later, a thermoelectric drill that would melt its way through the ice. Four cores were drilled under Bader's supervision in 1956 and 1957, two in Greenland and two in Antarctica, and the results were promising—though Bader thought at the time that they had already reached, at a depth of little more than 300 meters, the maximum length at which recovery of cores would still be feasible. Fortunately for the scientific understanding of climate, that guess turned out to be too pessimistic by a factor of ten.

Within decades, 3-kilometer-long cores would be drilled, a remarkable achievement that depended on the development of new drilling technology and analytic techniques. One of the first things to be puzzled out was the nature of the little gas bubbles found on the way down into the ice, for they contained clues to the earth's distant past. Why were the bubbles there? Basically, as new snow falls, air from the ambient atmosphere is trapped between ice crystals at the surface. As the snow is buried and starts turning to old snow, known as firn, the trapped air forms bubbles, which become quite spherical as the firn is compressed into ice—in Greenland's ice sheets, this characteristically happens at a depth of 57 to 70 meters. Ice drillers would discover, as they extracted cores deeper down, 900 to 1,300 meters from the surface, that as the ice "relaxed" at the surface, expanding as it warmed, the bubbles would reappear and snap, like ice cubes in a cool summer drink. Still farther down, at around 1,300 meters in Greenland, the air diffuses into the ice to form what scientists call a clathrate—molecules of one kind are completely trapped in the crystal lattice of another substance.

Langway's first important job for Bader was to determine the pressures of the bubbles, which required him to use a novel apparatus and exact procedures. He was successful, winning Bader's respect and affection. In 1961, SIPRE was folded into a new army lab being set up in Hanover, New Hampshire, the Cold Regions Research and Engineering Laboratory (CRREL). In due course, Langway was appointed chief of the snow and ice branch, having in the meantime finished his doctoral studies at the University of Michigan. At CRREL, with great efficiency and acumen, he would organize collaborations between U.S. and European scientists, which, by the end of the decade, would produce some startling revelations about the earth's prehistoric climates. One was the discovery of what scientists would call abrupt, rapid, or even catastrophic climate change. Even more important, there would be what Langway and the Swiss physicist Hans Oeschger called "the astonishing revelation that the carbon dioxide/air ratio varied contemporaneously with the climatic conditions experienced during glacial/interglacial shifts."[5] What the scientists discovered was that, when the earth moved into and out of its three or four most recent ice ages, as global mean temperatures rose and dropped, carbon dioxide levels climbed and fell too, in lockstep.

As a matter of pure theory, it's been understood since the mid-nineteenth century that trapping of the sun's radiant energy by the earth's atmosphere—strictly speaking, trapping of radiation that would otherwise reradiate from earth back into space—is what boosts temperatures on this planet enough to support advanced life.[6] And for more than a hundred years, the common belief has been that this greenhouse effect, as it's popularly, if slightly inaccurately known, would be strengthened by combustion of fossil fuels, because of the added carbon dioxide pumped into the atmosphere.[7] But only in the last few decades have those notions been proven as hard empirical fact, as scientists like Ramanathan directly measured flows of energy into and out of the atmosphere, others gauged present-day changes in greenhouse gas levels, and—most tellingly of all—a growing number of exceptionally enterprising individuals began to extract the history of the world's climate from a wide variety of sources.

Throughout earth's history in the narrow sense, the very recent period for which written records exist, climate trends could be gleaned from sources like the all-India monsoon record going back to the nineteenth century, and records from China of droughts, floods, and cold snaps going back to the first century A.D. (two millennia before the present or "B.P.," to use the language favored by professional paleoclimatologists). The Hudson's Bay Company required its captains and agents to record daily weather, including first snows and thaws, throughout the eighteenth and nineteenth centuries. A record going still further back comes from Spanish "rogation" ceremonies performed to bring rain or to end a deluge.[8]

Seeking clues to more ancient climates, students of the more recent prehistoric centuries have relied primarily on tree rings, from both living and preserved trunks. By matching the patterns in living trees with overlapping patterns in dead trees found in places like the embankments of the Rhine River and its tributaries, scientists can obtain information about rainfall and temperatures going back as much as 12,000 years. (The trick is to find situations where either temperature or rainfall is believed to have stayed fairly constant, so that the rings accurately indicate changes only in the other parameter.) Just as ingeniously, scientists learned to extract cores from lake beds and distinguish annual accumulations of sediment, and take various kinds of readings from what was found in each layer. In some regions, for example, pollen records showing how wildflowers had migrated in response to changing climate could be constructed for tens of thousands of years. Similar information could be obtained from animal middens, excrement containing residues of organic matter the creatures had ingested under different circumstances. From all such data sets covering the last 1,000 years, both the written histories and the physical indicators, a sharp rise in the world's temperatures in recent decades has been well documented.[9]

Even more dramatic, however, have been the discoveries made by those drilling into ocean beds and into polar ice. Their insights, especially those published in just the last two decades, have transformed our understanding of the earth's last half million years. From the character of the ice accumulated in Greenland and Antarctica and from the gases trapped in each layer, it's become evident that the earth's climate is not merely unstable, but violently unstable. Changes previously believed to have unfolded over thousands or tens of thousands of years were found to have taken place in less than a hundred, sometimes in less than ten. An intimate relationship between temperature and greenhouse gas levels, long assumed, has been demonstrated beyond doubt.

The story of these epochal discoveries is one of adventurous and imaginative individuals turning all manner of very new technology to the study of very old climates. Without the breakthroughs made during the 1930s in atomic chemistry, and without instruments developed in World War II, these pioneering scientists could not have done their work. Nor, most likely, could the job of exploring the world's most frigid regions have been accomplished without the mad tensions and pressures of the Cold War, which gave rise to opportunities that a handful of farsighted knowledge seekers were quick to seize.

The story is perhaps best begun before Langway entered the scene, in 1947, when a young Danish geophysicist, Willi Dansgaard, was assigned to help take geomagnetic readings at a weather station in northwest Greenland. There, Dansgaard and his young wife were "bitten with Greenland for life…its forces, its bounty, its cruelty, and above all its beauty," as he would later put it.[10] During a follow-on stint with Denmark's weather service, Dansgaard began to feel that earth's climate was a lot more interesting than its magnetism, an important foundation stone for what would be an unusual career.

When Dansgaard returned in 1951 to the University of Copenhagen's Biophysical Laboratory, where he had earned his academic degrees, he was put to work on instrumentation. His first job was to install a brand-new mass spectrometer. Basically, a mass spectrograph or spectrometer spews out a stream of charged matter across a magnetic field, so that the material can be weighed and identified according to the extent it is deflected in the field. The instrument had been invented by Francis William Aston at Cambridge University in 1918–19, and it immediately played a key role in the development of atomic theory, the branch of physics called quantum mechanics. One use of the mass spectrometer, on a very large scale, was in the Manhattan Project, to separate fissile uranium 235 for atomic bombs from the much more prevalent nonfissile U-238 isotope.[11] (Isotopes are variants of an element that are chemically identical but have different numbers of neutrons and therefore different atomic weights and numbers.) At the Copenhagen lab, the application in mind was the production and use of stable isotopes (isotopes that do not decay radioactively) for medical and biological research.

Because Dansgaard had been exposed to aspects of meteorology, unusual climates, and nuclear science, he happened to learn that rainwater contains lighter and heavier components: two variants of regular H_2O, the dominant one containing the O^{16} isotope of oxygen and a much rarer one containing the heavier O^{18}; and so-called heavy water, HDO, in which one of the hydrogen atoms is the heavier isotope called deuterium.[12] This nugget of information led to a startling discovery of great import. During a huge rainstorm in northern Europe during the weekend of June 21, 1952, Dansgaard wondered whether the isotopic composition of the precipitation might change from one shower to the next. To find out, he put funnels in beer bottles out on his lawn to collect water. He was the right man wondering at the right time, for "it turned out to be an unusually well developed front system," as he later recalled.[13] "When the rain began in western Jutland, it had not stopped raining in Wales 1000 kilometers to the west. I have not seen anything like it, at no time before or after. The miracle consisted in my starting the collection accidentally under these unusually favorable conditions."

Upon analyzing the water samples at the Copenhagen lab, using the mass spectrometer he had helped install and improve, Dansgaard discovered that the isotopic composition of the rainwater did in fact change, and that it changed in a regular way as the storm progressed. Specifically, because the heavier water molecules are less likely to evaporate from a surface but more likely to condense from a cloud, the higher and colder a cloud is, the less likely it is to contain the heavier waters. To look at the situation dynamically, the first rainfall will leave a cloud depleted in H_2O^{18} and HDO because of their greater propensity to condense. As the cloud rises and cools, still more of the heavier waters condense out, and so on. The bottom line—disarmingly simple—is that the temperature of a cloud from which rainwater falls can be inferred from the isotopic composition of the rainwater.

Having learned that, Dansgaard, a very determined perfectionist, according to lifelong associates, took the next logical step: to find out whether this relationship held true for rainwater falling in different parts of the world from clouds of widely varying temperatures. Through the good offices of the Danish East India Company, Dansgaard obtained enough water samples to convince himself and others that rain falling in temperate and polar regions had lower concentrations of the heavier isotopes than rain falling in the tropics. In getting the samples he needed to document this finding, he owed a good deal to the clubby character of Danish science, politics, and business. Copenhagen was a small world in which all the elites mixed comfortably and casually, and everybody interesting or important knew everybody else.

Through such connections, the lucky Dansgaard would now benefit from an even bigger favor. In the late 1950s, the World Meteorological Organization in Rome and the International Atomic Energy Agency in Vienna launched a Global Precipitation Network to gather samples from all over the world on a regular basis. The main objective was to track radiation from bomb tests or nuclear power plant accidents, which concerned the WMO as the world's designated monitor of the earth and the IAEA as its monitor of nuclear standards and safeguards.[14] Through diplomatic and science agency contacts in Copenhagen, Dansgaard was able to get access to the samples. The key intermediary was a high official at the Danish Atomic Energy Commission who, during the war, had made a name for himself channeling treasury department funds to the Danish Resistance and getting food to Nazi concentration camp inmates. Analysis of the samples enabled Dansgaard to definitively establish, on a global basis, the direct correspondence of cloud temperatures and isotope ratios. The result was his landmark paper, "Stable Isotopes in Precipitation," which the journal *Tellus* published in 1964, and which continues to be cited frequently in scholarly references.[15]

What made that paper so immensely important was its connection to another insight Dansgaard had already had a decade earlier. Because of his experience in Greenland and his musing about what had happened to the polar ice cap in past geologic ages, it had occurred to him that isotope ratios might be the key to learning the temperature of the world's atmosphere in epochs gone by.[16] "I was immediately sure it was a good idea," he would later say, "maybe the only really good one I ever got."[17] It was in fact a brilliant idea, and for the rest of his working life, with growing crews of oarsmen at his beck and call, Dansgaard would pursue it with the obsessiveness of a Viking raider.

Two years after the *Tellus* paper appeared, Dansgaard obtained a coveted professorship at the University of Copenhagen, on the basis of the doctoral dissertation he had completed in 1961. It summed up results from a sailing expedition he had made to Greenland in 1958 to collect samples from icebergs, to further test his ideas about oxygen isotopes and temperatures. In 1964, when he had another opportunity to visit Greenland with fellow scientists—this time to take samples for an attempt to date ice by means of the radioactive decay process in silicon 32—he found something astonishing, an enormously elaborate U.S. base at Camp Century, 220 kilometers east of Thule in the far northwest corner of Greenland. Thule itself was a major NATO air base, with thousands of military personnel, jet fighters, and the C-707 transport planes used to refuel B-52 strategic bombers in the air, Langway recalls. The subsidiary base at Camp Century, pretty literally carved out of the ice, was one of the more bizarre by-products of the Cold War, but one that proved crucial to the fast-evolving sciences of glaciology and paleoclimatology.

It was operated by CRREL, the lab in Hanover where Langway was doing basic environmental and ice research, but which was mainly dedicated to the development of equipment and techniques for military operations in frigid conditions. That mission had taken on new life with the Cold War, first of all because of a general feeling among American generals that the Red Army knew a lot more than they did about fighting on ice and permafrost. But more specifically, the generals were eyeing Greenland as a particularly favorable location for U.S. systems to provide early warning of a Soviet missile attack, and possibly even as a site for forward deployment of U.S. ballistic missiles. Under these circumstances, Denmark's colonization of Greenland was an advantage for Danish scientists seeking to piggyback research projects on U.S. military operations. Technically, the United States could not do anything in Greenland without the permission of the Danish government,[18] and, while relations between the two countries were very friendly, proposed U.S. military activities in Greenland were potentially controversial in Europe, and therefore required not just pro forma permission but strict oversight from the Danish authorities.

Dansgaard and his colleagues found at Camp Century a whole town built under the ice, with a main street, mess halls, post exchange ("PX") stores, and recreational facilities. People, goods, and equipment were ferried around by gigantic custom-built D8 tractors and trains: the tractors, with 1.5-meter-wide treads, pulled 8 wagons with 3-meter-diameter balloonlike tires. The base was powered by its own nuclear reactor, and the remains of a sub-ice iron railway lay in tunnels, eventually to be bent out of shape by shifting ice floes.

Later Dansgaard would be told and would believe, perhaps too credulously (Langway thinks), that Camp Century had been built as part of a highly classified project called Ice Worm to explore the idea of installing U.S. ballistic missiles on mobile launchers in the ice tunnels. If there's anything to that, it would cast the events leading up to the Cuban missile crisis, and the deal that ended the crisis by trading U.S. missiles in Turkey for the Soviet missiles in Cuba, in a new light. However that may be, at the time, Dansgaard and his fellow Danes came away with an impression of their American friends as having rather too many dollars and perhaps too few of what Agatha Christie's detective Hercule Poirot famously called those little gray cells.

What interested Dansgaard most at Camp Century was a core drilling derrick installed in a trench under the snow. Though he couldn't divine exactly why or for what purpose it had been built, he thought immediately of how ideally suited it would be for taking ice samples to do oxygen isotope analysis. He learned that the drill had been designed and built by CRREL's B. Lyle Hansen, the first major inventor of deep-ice coring equipment. The Americans would use that drill two years later in 1966, to go all the way down through the ice to bedrock at Camp Century and take the first very long core. Dansgaard's Copenhagen lab would provide crucial help with the stable isotope analysis from that core and its successors to determine atmospheric temperatures that had prevailed over Greenland for tens of thousands of years.

From conversations with the veterans of the early ice coring expeditions, it's not easy to fathom what the pioneers had in mind when they first began to poke around in the polar ice sheets in the 1950s and 1960s. Looking back now, it almost seems that ice science was like the proverbial mountain that had to be climbed simply because it was there. Large military transports were available for ferrying equipment to remote ice sheets, while smaller planes could take scientists to outposts and rescue them if they got into trouble. Newly developed vehicles like the huge Caterpillar wagon trains at Camp Century could move people around on the ground.

The most important piece of equipment in the new science of ice coring was of course the drill itself. Hansen's first models included a thermally heated augur and a pretty straightforward electromechanical drill derived from ones developed for oil exploration and extraction. The latter consisted of drilling blades at the end of a rotating steel cylinder, with a compressed-air system bringing shavings to the surface.[19] It soon became apparent that the early models had to be improved and optimized in a number of ways for practical removal of cores, especially as bore holes went longer and deeper. Techniques had to be devised for removing core segments while drilling continued, fast enough to make the whole drilling operation economically practical and humanly tolerable. The drill had to cut as fast as possible, but not so fast it overheated and got stuck deep in a hole, bringing operations to a halt—sometimes irreversibly. As the drill descended, the space above it had to be filled with a fluid to equalize pressure in the hole and prevent it from freezing over the drill. As time went on, special operations such as drilling through relatively warm ice or in remote mountain glaciers required development of much more compact and energy-efficient drills. One such type, a thermal drill, has a heating element at the tip rather than at the blades, so that the drill simply melts its way through the ice, with a system to remove the meltwater to the top.

As it dawned on scientists that ice cores might go a lot deeper and would be much more valuable as they went much farther back in time than had been guessed initially, it became crucial to drill in just the right spot. This meant finding locations where there was good reason to think the ice had stayed frozen and undisturbed for tens or hundreds of thousands of years, and where little or no ice drifting had occurred, so that the cores extracted would be truly sequential, year by year. Naturally the goal was to go all the way to bedrock, so careful radar surveys from planes were necessary to determine whether and where that might be possible.

Development of advanced instrumentation also was required. When the first relatively shallow holes were bored, dating ice did not seem especially challenging, as layers close to the surface were distinguishable to the eye. But when the cores went deeper, where annual layers were no longer visible, figuring out where one year ended and the next began required an array of highly sophisticated instruments. One such device was the mass spectrometer, which sometimes could be used not only to gauge changes in temperature from year to year but also to detect seasonal variations with each year's water isotopes. Langway played a key role in establishing the feasibility of distinguishing one year from another, within certain ranges, by using the spectrometer to detect seasonal differences in oxygen isotope levels.

A novel instrument invented and employed in the 1970s was the Coulter counter. It relies on lasers to detect dust particles in melted water from samples that are pumped through capillary tubes. Since the air over Greenland is usually driest and dustiest in the spring, when particles blow in from as far away as the Himalayan highlands, the Coulter counter can precisely discern each summer's point of maximum warmth and dryness.

Another instrument, relatively simple in principle and designed to work automatically, continuously, and fast, measures the acidity of ice. The core is split, two electrodes are dragged along the flat surface, and spikes in electrical conductivity between the electrodes indicate the presence of higher acidity, particularly sulfuric acid from volcanic eruptions. This device, invented by a Dane, Claus Hammer, enables scientists to detect fallout from historically dated volcanic eruptions in the immediate past millennium and to calibrate other dating techniques to match those records. (Discovery of the volcanic residues, which were found to have caused global cooling when fallout was in the air, provided support for "nuclear winter" scenarios that were developed in the 1980s. Models indicated that an all-out nuclear war, besides killing hundreds of millions of people instantly, would also produce a devastating global chill.)

Even with all the new technology and infrastructure, and a political environment that favored ice core research, somebody had to have the idea that interesting things could be learned from drilling into ice to really get that research going. In the United States, Bader was the catalyst, and Langway, his protégé, was quick to identify the other most highly qualified pioneers. It was he who would broker a marriage between the little community of U.S. glaciologists and top European experts reared in the Scandinavian and Alpine traditions of ice study.

As the work from the 1957–58 geophysical year was digested, recalls Langway, "other interested scientists were struck that the cores revealed crucial information about the history of climate." The stage was set for the first really long core to be extracted. Langway found himself in charge of the analysis that would be done on the ice from the next big coring effort, which would be carried out by the United States, this time at Camp Century, again with a drill developed and perfected by the ingenious engineer Hansen. As the core was being extracted, Langway set about finding the very best people to analyze the ice. "Chester had a deep understanding of what was important—really key—and wanted the best possible science," recalls Sigfus Johnsen, an Icelandic disciple of Dansgaard whose drills eventually supplanted Hansen's.[20] One of Langway's major recruits was Dansgaard, who contacted him in 1966 after spotting the Camp Century drilling rig and then, at Langway's invitation, flew over to the United States to negotiate a working arrangement. Dansgaard took charge of the stable isotope analysis for the Langway–Hansen Camp Century core, and Langway came to appreciate him as a person "dead set" on getting a job done, and having "the scientific prowess to make a program fly."

Just as important was a Swiss recruit named Hans Oeschger, whom Langway met at a conference in Austria in 1962. His attention was caught by Oeschger's efforts to do carbon 14 dating in very small samples, samples in which the expected quantities of carbon 14 would be minuscule. Langway flew straight to Bern to get better acquainted. In due course, it became apparent that trying to estimate an ice sample's age from the radioactive decay of the carbon 14 isotope wouldn't work very well: the isotope made up too minute a fraction of the carbon dioxide trapped in tiny bubbles to yield meaningful readings. And any carbon dating would be confined only to the most recent 50,000 years or so, because almost all the carbon 14 in a given sample decays in less time than that. But what really mattered, despite these disappointments, was Oeschger's skill at measuring tiny quantities of carbon dioxide.

As a graduate student in the 1950s, Oeschger had invented and built a novel instrument called the proportional counter to detect and measure ultralow levels of radioactivity; unlike a Geiger counter, it was capable of identifying the type of radiation prompting a signal. Completed in 1955, it came to be known as the Oeschger counter, and was for many years a leading instrument in radiocarbon laboratories.[21] In 1959, using the device, Oeschger and two scientific collaborators were the first to radiocarbon date Pacific Ocean deep water, which helped set the stage for the systematic study of oceanic circulation patterns. Working with Langway in the 1960s and 1970s, Oeschger learned how carbon dioxide gets trapped in ice, how to extract it, and how to select ice samples with air bubbles retaining the composition they had when they were formed.

Langway, Oeschger, and Dansgaard made a powerful trio, with Langway interested in the physical and chemical properties of ice, Oeschger in carbon dioxide and radioactive isotope concentrations, and Dansgaard in the stable isotopes' temperature variations. In the next decades they produced what is arguably the most compelling and provocative evidence of large-scale and rapid climate change obtained from any source.

Dansgaard, Langway, and Oeschger (*left to right*).   *Source:* Photograph by J. Murray Mitchell, courtesy AIP Emilio Segré Visual Archives / Gift of Chester C. Langway Jr.

In 1961 Hansen took his drill to Camp Century, and by 1966, using a variety of drills, a 1.5-kilometer-long core had been obtained, revealing about 125,000 years of atmospheric and environmental history. Dansgaard's isotope analysis showed pronounced warming and cooling periods in times for which there are historical records, corresponding to well-documented episodes such as the Little Ice Age of medieval times and a short warming trend that culminated in the 1930s. Farther back, very sharp coolings and warmings were detected, corresponding to glaciation and deglaciation, notably the Allerod/Bolling event that terminated the last ice age. When Dansgaard and others presented their preliminary results at an international symposium in 1968, it caused "quite a stir," Dansgaard reported,[22] and resulted in his being invited on the spot to give a talk later that year at the Nobel Prize symposium in Uppsala, Sweden. The following year, Dansgaard and colleagues published their definitive summation, "One Thousand Centuries of Climate Records from Camp Century on the Greenland Ice Sheet," in *Science* magazine, the top U.S. journal.[23]

Building quickly on that success, Hansen took his equipment down to Antarctica and started to drill with a vengeance at the Byrd station, reaching bedrock 2,164 meters—more than 2 kilometers—down in 1969. Oxygen isotope analysis in that core produced pictures of the last ice age and the following deglaciation that were dramatically consistent with Camp Century's. Just as important, the ice at the Byrd station turned out to be much more pure than that obtained from the Camp Century core, which was found to be too contaminated with carbonates to yield reliable measures of carbon dioxide levels by means of Oeschger's techniques.

Subsequently the drillers returned to Greenland to take a core at Dye 3, the site of a big U.S. radar installation that was part of the famed DEW line early warning system.[24] This was part of what they called the Greenland Ice Sheet Program, or GISP, a U.S.–Danish–Swiss collaboration. Conveniently, the drillers were able to work underneath the framework of a giant radar building, installing their equipment in its shelter. (The structure was jacked up each year, as the underlying ice became compressed and snow drifted under the frame.) For analysis, the core was split down the middle and studied independently at the U.S. Army's Hanover lab and in Copenhagen. That core provided an opportunity to test a new thermal drill devised by Hansen, to strengthen collaborative procedures, and to improve analytic techniques. It was the first of twenty relatively short cores to be extracted from various Greenland sites over a ten-year period as part of the GISP.

This first Dye 3 core was only about 400 meters long and was taken, it seems, largely as a matter of convenience. The site in southern Greenland really did not satisfy the key conditions for a good core: no surface melting (because it could result in soluble gases contaminating the atmospheric traces left in each year's ice bubbles), little or no disturbance to deep stratification from ice flows, and good snow accumulation with little melting.[25] So the drillers prepared to extract a long core from a more selectively chosen site in central Greenland, called Summit. Meanwhile, however, the U.S. science authorities began to quibble about the high expense of going to a new site, and meanwhile, Hansen had run into difficulties with a new drill in Antarctica.

Dansgaard's younger colleague Sigfus Johnsen stepped into the breach with a new design. Some salient features of his Istuk drill (a contraction of the Danish word for ice and the Greenlandic word for spear or awl) were its much greater compactness, a tilting mechanism for easier core removal and maintenance, a novel method of directly extracting ice shavings rather than dissolving them as drilling proceeded, and a microchip built into the drill head to keep operators apprised of key parameters like speed, inclination, pressure, temperature, battery charge, and the mechanical resistance encountered by the cutting mechanisms. Johnsen also developed techniques to speed preparation of samples, a crucial economic consideration as drills went deeper, and to automate some sample analysis. By this time, recalls Langway, the Danes could process 256 samples in a night and be ready the following evening to process another day's yield—a record that's never been beat.

In 1977, the U.S. authorities agreed to join the Danes in a collaborative deep core back at Dye 3, chosen for logistical convenience and economy, despite the site's obvious drawbacks, using Johnsen's Istuk drill. They struck bedrock on August 11, 1981, a little more than 2 kilometers below the surface. Like the Camp Century core, the Dye 3 sample provided a strong hint that the periodicity of ice age glaciation and deglaciation was strongly associated with the Milankovitch cycles—the subtle variations in the Earth's relationship to the Sun (described at the end of chapter 2). And Dye 3 confirmed Camp Century's discovery of sharp oscillations in temperature roughly every 1,500 years—an observation that seemed almost too weird to credit when first made. With the cycles now showing up not just in one core but two, and subjected to close scrutiny by international teams, they began to get a lot of attention, and not just from glaciologists.

The general pattern in these dramatic oscillations, which came to be better known as Dansgaard–Oeschger cycles, was for them to begin with a very abrupt warming, on the order of 10 degrees Celsius on average, and to end with a much more gradual cooling. The warmings, which Dansgaard referred to in print[26] as "spectacular changes" in temperature, took place in as little as a century. Altogether, 24 such cycles were seen in the camp Century and Dye 3 cores, in parallel with one another.

The Dansgaard–Oeschger cycles, like the ice ages themselves, remain something of a mystery. Coming up with a completely convincing explanation of them is to this day one of the grand challenges in paleoclimatology. The most widely accepted account is one developed largely by the climatologist William Ruddiman, Oeschger, and Wallace Broecker, a geochemist at Columbia University's Lamont-Doherty Earth Observatory, whose ideas will be discussed at greater length in chapter 7.[27] It rests on presumed changes in ocean circulation, a notion that Ruddiman may have been the first to enunciate,[28] and which Broecker championed. It distinguishes between two types of millennial cycle, the usual Dansgaard–Oeschger events and a category of more drastic but rare cycles known as Heinrich–Bond events.

The German scientist Hartmut Heinrich was a specialist on debris that had collected at the base of glaciers during the last 100,000 years, to be "rafted" out to sea in icebergs and finally deposited on the seabed far from its origins, in a series of layers. Gerald Bond, a colleague of Broecker's at Lamont, had made observations and reached conclusions from ocean bed cores that provided a key ingredient of the provisional explanation for the Dansgaard–Oeschger cycles. Though merely a hypothesis, the theory is so suggestive and disconcerting that it bears some attention. Bond noticed a pattern within the Dansgaard–Oeschger oscillations: several in succession would get colder and colder at their most extreme, and then—just before a really abrupt warming—a lot of iceberg-rafted debris would show up in ocean sediments. Bond, Oeschger, and Broecker hypothesized that as the Dansgaard–Oeschger cycles got colder, Hudson Bay would fill up entirely with ice, but when the ice reached a certain depth, it would act as a thermal blanket. Heat from the earth's interior would eventually melt the bottom of the ice sheet, and suddenly the whole sheet would give way, slide out into the North Atlantic, and break up, carrying accumulated debris out into the more southerly waters. But as the icebergs melted, an infusion of fresh water would prevent the normally salty surface waters of the North Atlantic from sinking, interrupt the normal flow of warm southerly waters into the North Atlantic, and lead to a gradual cooling. If that infusion of water took place only in the far north, the result would be a normal Dansgaard–Oeschger cycle, but if it was more widespread and affected waters in intermediate latitudes as well, there would be a more pervasive and severe "Heinrich" event.

A wide variety of oceanic evidence indicates that the regular Dansgaard–Oeschger cycles would make themselves strongly felt throughout most of the Northern Hemisphere, down to the equator and sometimes across it. The double-impact Heinrich events would affect the whole globe more perceptibly. "Comparison of the Greenland and Antarctic cores showed," science historian Spencer Weart has observed,[29] that at least some of "the climate changes were truly global, coming at essentially the same time in hemispheres."

Before Camp Century and Dye 3, sediments bored out of ocean floors had provided the most insight into earth's climatic prehistory. During a 1947 Swedish expedition, scientists developed a way to extract a column of sediment from a bore hole in the ocean floor without disturbing the sequence of accumulated sediments, and in the next decade, radiocarbon dating was used to calibrate the rate of sedimentation.[30] By the mid-1960s, several drilling expeditions around the world had produced results going back several hundred thousand years, showing a correspondence with the Milankovitch cycles and indicating that the world must have gone through dozens of glaciations and deglaciations—not just four, as nineteenth-century geologists had come to believe. It was becoming apparent that the modern era in which human civilization had emerged, the Holocene, was a blessedly benign anomaly in the much larger scheme of things.

Once the results from Camp Century and Dye 3 were in, however, it was generally recognized that ice cores now provided the most authoritative record of past climates. For all the qualms about the Dye 3 site, two long cores were much better than just one, and the consistency of the two records was striking. The rather sensational findings set off what a writer for *The New Yorker* magazine aptly called an "ice rush,"[31] with the ironic consequence that the spirit of international cooperation that had prevailed up until then quickly broke down. Now that coring was recognized as a really serious business, the U.S. National Science Foundation began to "micromanage" it, says Langway, and that seems to have undermined the crossnational relations that had evolved over two decades. Dansgaard and Johnsen got the impression that the U.S. authorities felt the Danes had been enjoying a "free ride," though it was Johnsen's "ice spear" that had produced Dye 3 and Dansgaard's Copenhagen lab that had developed the key analytic techniques and performed much of the actual analysis. (Dansgaard had a reputation for being cantankerous and difficult to work with. But he had done analytic work on all the major Greenland cores so far "without ever charging a penny for it," notes Langway.) Perhaps, taking a longer view, the collapse of the fruitful U.S.–European collaboration was just an early signal—weirdly analogous to other early warning signs from the Greenland sheets—of the larger deterioration in U.S.–European relations that began with the Cold War's thawing and end.

The U.S. and European teams went their separate ways after Dye 3, each drilling a new long core at Summit sites in north-central Green-

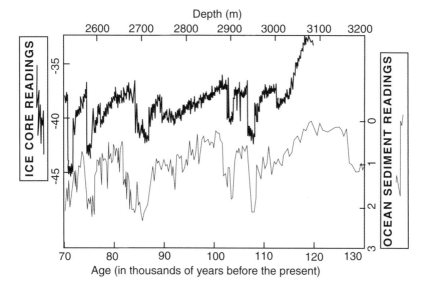

Consistent readings from Greenland ice and Atlantic ocean sediments
Source: Dansgaard, *Frozen Annals*, fig. 14.8

land in the 1980s and 1990s, separated by about 30 kilometers (one called GRIP, or Greenland Ice core Project, the other GISP 2). Those cores were better situated than Dye 3, and from a scientific point of view, there was something to be said for doing two long cores completely independently. So when the results came in and were processed in the early years of this century, they gave still greater credibility to the earlier discoveries and provided more detailed data on ice age onset and termination, as well as on the Dansgaard–Oeschger and Heinrich–Bond cycles. The results also raised a host of new issues about what was going on globally and how events in the far north squared with what was going on in the south—in Antarctica especially, but also in the equatorial regions. Even in the nearsighted perspective of historical time, the cores showed some nice consistencies with well-documented events: GISP 2, for example, gave a snapshot of fluctuations in which important events could be discerned, like a European period of torrential rains and famine from 1308 to 1318 and a period of North Sea storminess from 1343 to 1362, known as "the great drowning."[32]

Particularly dramatic was the correspondence between temperature readings from Europe's NorthGRIP core and those from ocean borings done off the coast of Portugal (see graph). In 1976, a comprehensive study of ocean sediment samples definitively established temperature fluctuations corresponding to the Milankovitch periods of 23,000, 41,000, and 100,000 years.[33] As ocean sediment and ice samples continued to yield closely consistent findings, it became unmistakably clear that ice ages—many more than just the four postulated in the nineteenth century—had occurred at the 100,000-year Milankovitch beat, with remarkably small changes in solar radiation mysteriously producing huge and disconcerting changes in the earth's climate.

Yet even as the story of polar ice research approached a kind of crescendo and climax, the United States allowed its leadership to slip, first technologically, with the replacement of Hansen's drilling rigs by Johnsen's, then scientifically, as the Europeans worked more exclusively with one another, and increasingly with Russian drillers and scientists as well. In the last decade of the twentieth century, the initiative increasingly was taken by drillers working the South Pole sheets, notably groups of collaborating French and Russians, who during the Cold War had been looking on jealously from the sidelines as ice-based paleo-climatology got exciting. The Russians began to poke around their Antarctic station in Vostok, though initially they couldn't seem to get their equipment to function. A French team based at Grenoble started to work closely with the Russians. Its scientific leader, Claude Lorius, was known as a personable man eager to collaborate with everybody, but without seeming to have—at least initially—anything unique to put on the table. Eventually, though, in something like a tortoise-and-hare tale, the Franco-Russian group drilled a very long core at Vostok and produced a compelling picture of earth's climate during the last 430,000 years.

When Oeschger first was turning his attention to the carbon dioxide trapped in ice, as noted, it had turned out that radiocarbon dating was really not practical for long-term ice dating. Then, after enormous work extracting and measuring the carbon dioxide content in the first long Greenland cores, many of the results proved almost worthless when it was discovered that much of the ice had been contaminated by dust containing carbonates. Nevertheless, in the course of studying those cores, Oeschger and his collaborators gradually perfected the art of carbon analysis, and during the 1980s and 1990s they established a close correspondence between temperature and carbon fluctuations.

Just as importantly, Oeschger firmly established how much carbon dioxide was in the atmosphere before the industrial revolution began, so that increases in carbon dioxide levels could be charted for the last 250 years.

As early as 1980, Oeschger and colleagues published an analysis in the journal *Radiocarbon* of the carbon dioxide in air bubbles in the Byrd and Camp Century cores,[34] and found that concentrations were significantly lower during the last glaciation than afterward, possibly as much as 50 percent lower. A paper published by the Swiss team in the top British journal *Nature* in 1982, based on an analysis of 3 cores, found that increases in atmospheric carbon dioxide from the last ice age to the current Holocene era ranged by factors of 1.2 to 1.4.[35] A 1988 Swiss paper considering "pitfalls" in carbon dioxide study concluded, from further analysis of the long Byrd core, that glacial carbon dioxide levels were about 30 percent lower than the Holocene, preindustrial levels.[36] Specifically, the glacial carbon dioxide concentration fluctuated between 180 and 200 parts per million (ppm), while the preindustrial Holocene level was around 280 ppm.

"The discovery of natural oscillations in greenhouse gases from fossil air trapped in polar ice ranks as one of the most important advances in the field of climate and earth science," wrote the noted geologist Thomas M. Cronin at the time,[37] hailing the Swiss and French achievement. Determination of the preindustrial carbon dioxide level, crucial to evaluating global warming in the industrial era, also was a feat of no small import. Since 1957–58, at the behest of Roger Revelle and Hans Suess at the Scripps Institution of Oceanography in La Jolla, California, direct measurements of carbon dioxide concentrations in the atmosphere were made atop Mount Mauna Loa in Hawaii. This ongoing record of steadily rising concentrations, done by their colleague C. D. Keeling, has come to be Exhibit A in the debate over global warming and the canonical presumptive proof that human activity must be affecting global climate. But the record had an annoying shortcoming in that it started out in midstream, long after the industrial revolution began in eighteenth-century Britain.

Oeschger proposed now to connect the dots. Relying on the best data they had obtained on carbon concentrations in ice cores, Oeschger and his colleagues estimated in 1985 that the atmospheric carbon dioxide concentration in the year 1750 was about 280 ppm, and had increased in the meantime by 22.5 percent, to 345 ppm in 1984.[38] Twenty years later, in 2004, the level of carbon dioxide in the atmosphere was 377 ppm, 35 percent higher than in 1750.

From determination of the variations in carbon dioxide levels and the determination of the preindustrial level, it was a short step to finding exact correspondences between carbon dioxide levels and global temperatures. In October 1987, *Nature* published three papers from the Franco-Russian group drilling at Vostok. It is the coldest station on earth, 1,400 kilometers from the nearest coast, with a mean yearly temperature of minus 70 °C. The three 1987 reports delivered what the magazine's expert commentator, Eric T. Sundquist of the U.S. Geological Survey, called "remarkably close association between CO_2 and climate variation, extending through the last interglacial period," proving beyond doubt "that the global climate system and carbon cycle are intensely interactive." At the same time, he said, though both the deuterium isotopes and carbon dioxide records squared fairly well with the 40,000- and 200,000-year Milankovitch cycles, the connections were not simple, and there were some puzzling discrepancies between these records and others.[39]

Additional drilling at Vostok soon showed that the world's present greenhouse gas levels are unprecedented for the last 420,000 years and that carbon and temperature levels are inextricably linked in a lockstep relationship. On June 3, 1999, *Nature* published an article based on the Vostok cores producing what its expert reviewer, the Bern climatologist Bernhard Stauffer, called a "cornucopia of ice core results."[40] Drilling had started in 1990, with U.S. scientists joining the Franco-Russian team, and reached bedrock in 1998 at 3.6 kilometers. The article summarizing the results, compiled mainly by the Grenoble group, contained what is surely one of the most striking charts in the history of climate research, and perhaps the most striking of all. It shows that if changes in greenhouse gases and temperature variations are plotted against time, the correspondence is so close that the two sets of data could almost have been plotted on the same lines. The right axis—which might reasonably be dubbed the Dansgaard axis— records stable isotope levels; the left axis—the Oeschger axis, if you will—records levels of carbon dioxide and methane. The four peaks and troughs so vividly seen represent four transitions from glacial warm epochs, starting at 335,000 years B.P., 245,000, 145,000, and 18,800. The roughly 100,000-year periodicity, along with the shorter and subtler cycles, correlates well with the Milankovitch cycles, sup-

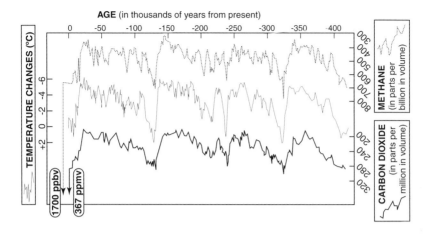

The Lockstep Relationship of Greenhouse Gases and Temperature
Source: Lorius

porting the idea that "changes of the orbital parameters of the earth (eccentricity, obliquity, and precession of axis) cause variations in the intensity and distribution of solar radiation, which in turn trigger natural climate changes," as Stauffer put it.

It remains unclear why the correspondences between carbon dioxide and temperature levels in the 420-millennium record are so exact, and what causal sequences are at work. If surprisingly subtle changes in solar irradiation are prompting sharp temperature changes, are the changes in carbon dioxide levels merely effects of the temperature changes? Or do the temperature changes somehow induce changes in carbon dioxide levels, which in turn amplify the temperature changes, in a chain reaction? And what if the causal chain is reversed, with carbon dioxide changes being the initiating events? Will we then continue to see the same lockstep relationship between carbon dioxide and temperature levels?

That the answers to such questions are not known is no ground for complacency. If scientists could say precisely how greenhouse gases and temperatures have interacted in past eras, then, looking ahead, perhaps they could say with confidence that as greenhouse gas levels rise sharply in the next century, drastic consequences can be excluded. But in fact, all they know is that in the past, sharp declines in greenhouse gas levels were associated with cataclysmic events, and that in the future, equally sharp increases in greenhouse gas levels will occur unless the world goes about its business rather differently.

The Vostok core represented, in a sense, the climax in a 50-year saga. In recognition of the path-breaking work begun by Bader and Langway and culminating in Lorius's Vostok core, Dansgaard, Oeschger, and Lorius were honored in 1996 with the Tyler Prize—the most prestigious award in the field of environmental science.[41] But Vostok is not the end of the story. An even longer Antarctic core, drilled at the so-called Dome C site, has in the meantime pushed the climate record back 740,000 years. Scientists studying the Dome C core noticed striking similarities between the transition that occurred 430,000 years ago ("Termination V"), which was unusually long, and the interglacial transition we are in now, prompting them to speculate that "without human intervention, a climate similar to the present one would extend well into the future."[42] It would seem, accordingly, that we do not have greenhouse gases to thank for the fact that we are not already entering another ice age—a train of thought that has sometimes tempted global warming skeptics.

What is especially disturbing about the Vostok and Dome C reports, even taking into consideration all the uncertainties about leads and lags and causes and effects,[43] is the scale of the greenhouse gas changes, compared to what's happening at present. The changes in carbon dioxide levels of 80 to 90 ppm between glacial and interglacial epochs are of the same order as the increase in carbon dioxide registered *so far* since the beginning of the modern industrial era. As this century progresses, unless concerted remedial action is taken, contemporary increases in greenhouse gases will soon far exceed the increases or decreases that accompanied the world into and out of the four most recent ice ages. By the same token, the temperature changes between glacial and interglacial periods—about 9 degrees Fahrenheit or 5 degrees Celsius—are similar in scale (though opposite in sign) to the changes that could occur in this century if carbon dioxide levels double.

Estimates of how much temperatures could change are derived from highly sophisticated computer models. But the startling discoveries of the ice drillers and analysts have presented climate modelers with a big challenge, and the jury is out on whether they will be able to meet it. As Dansgaard says on the last page of his memoir, *Frozen Annals*, combining all the various feedbacks between ocean and atmosphere into a convincing model able to predict future climate is a task of "overwhelming complexity." And while huge resources and tremendous brainpower are being brought to bear on that task, "one cannot even be sure that the climate is predictable at all."[44] That said, computer modeling is a science that also has made revolutionary strides in the last generation, and it may yet rise to the challenges posed by the ice research community.

CHAPTER 6

The Modelers

IN JANUARY 2005, when the American Meteorological Society convened in San Diego for its annual meeting, there were two special symposia honoring giants in computerized climate modeling. Fittingly, one of the all-day panels paid tribute to a scientist generally associated with fundamental limits to the predictability of weather, while the second recognized a modeler who has done perhaps more than any other person to make climate predictable.

One scientist, Edward Lorenz, an elderly meteorologist at the Massachusetts Institute of Technology, is very nearly a household name, at least among the mathematically minded. During the 1950s, Lorenz made a startling discovery in experiments with an early, rather primitive weather model.[1] After setting up a somewhat realistic opening situation, with some basic atmospheric data and some equations thought to describe the relationships among key variables, he fed it all into a simple computer and let the machine rip to see what would happen over time—the basic procedure that has become routine in climate modeling. But when he took results, fed them back into the computer, and reran them, as a check, he found the computer giving him clearly different final outcomes. At first Lorenz thought there had to be some kind of computer malfunction at work. But after retesting and rerunning his model many, many times, he discovered that the highly unpredictable differences in final outcomes resulted from very small rounding differences he had created when reentering data.

That discovery led Lorenz to formulate a fundamental principle in what has come to be known as chaos theory, a branch of mathematics that deals with interacting nonlinear equations, equations in which dependent variables are not a simple, direct function of independent variables but, rather, have relations that change with different values of the fundamental variable. The principle formulated by Lorenz is that tiny changes in initial conditions can lead, in time, to essentially unforeseeable changes in final outcomes. In what would be the canonical popular statement of the principle, he would ask in 1972 whether the flapping of a butterfly's wings in Brazil can set off a tornado in Texas—a question that he answered, essentially, in the affirmative.

The exciting story of Lorenz's discovery has been ably told by the science writer James Gleick in the opening chapter of his much-admired book about the origins and character of chaos theory (*Chaos: Making of a New Science*) and therefore needs no further detail here. But the other scientist honored with Lorenz at the meteorological society meeting in January 2005, Syukuro Manabe, is not a household name except in the rather small community of climate modelers, where he is legendary. His story, in its own way as absorbing as Lorenz's, deserves to be much better known.

It was Manabe who created, with colleagues, the first credible computerized model of the earth's atmosphere, properly visualizing the role of the greenhouse gases, and the first "coupled" model linking basic processes in the atmosphere to processes at the earth's surface, and to the world's oceans. In the course of doing that, Manabe—whose interest was purely scientific, not environmental or political—showed just how emissions from fossil-fuel combustion were making the planet hotter. Just as importantly, Manabe and his scientific companions developed detailed images of how growth or retreat of ice sheets affected climate, enabling them to run simulations testing ideas about the Dansgaard–Oeschger cycles described in the previous chapter, as well as some ideas about what accounted for ice sheets' advances in ice ages and subsequent retreats. Not least, Manabe used his models incorporating air, sea, and ice to test what would happen if carbon dioxide levels in the atmosphere doubled or quadrupled, versus what might have happened when they dropped by half, the way they did in the most recent ice age, per Oeschger.

That is not all. Manabe also became known not just for his specific experiments and experimental results but also for a general philosophy of modeling, which would have just as much influence as his discoveries. In a nutshell, he is a stickler for simplicity and model consistency, not only because errors and uncertainties inevitably multiply as a model gets more complicated but also because with added complexity, it becomes almost impossible to tell what the model is actually doing. For Manabe, a model is an aid to physical understanding, not a substitute for it. This means that data fed into a model must be solid and sensibly selected—otherwise it's "garbage in, garbage out," as he never tires of saying. And the equations used to represent the presumed interactions among variables must be sound physics, not just guesses about what might be going on. If those fundamentals are in order, then what comes out should produce not just forecasts but new insight into interrelationships in the climate system.

The basic Manabe procedure is to feed somewhat arbitrary data and the appropriate equations from classical physics into a model, run it, pause, see how well it's reproducing real processes, feed improved data and knowledge into the model, run it again, pause again, evaluate again, and so on. A small but highly dynamic man now in his mid-seventies, Manabe will turn sideways when describing that basic procedure, taking a wide dance step to the left as he faces his interlocutor, draw his feet together and pause, then take another dance step to the left, pause again, and so on. Charmingly, Manabe somehow combines a ballet dancer's natural grace—using his own body to convey his inner ideas—with an absentminded professor's clumsiness and inattention to externals. In conversation with a taller person, if he gets excited (as he almost inevitably will), his feet will suddenly rise off the ground, and for an astonishingly long balletistic moment, all four of his limbs will be in simultaneous motion, his feet and hands physically describing the independent movements of the four variables he happens to be contemplating. So intense is this instinct for concrete representation of abstract ideas that his children, when they were growing up in Princeton, used to tell friends that their father did "show and tell" for a living. When he'd leave for work in the morning, a former student and protégé reports, his wife, Noko, would remind him to "try not to run into any trees today." Colleagues following him in cars would report that the experience was like watching the beloved animated cartoon figure Mr. McGoo, the very nearly blind fellow who somehow always narrowly escapes catastrophe, despite an almost total obliviousness to his physical surroundings.[2] Such characteristics have endeared "Suki" to his many students and lifelong collaborators, and made him not just a model modeler but an embodiment, too, of what a scientist-*mensch* should be. They speak not just of his insistence on "digging in" to fully understand what a model's doing but of his generosity—with his time, with the fruits of his labors—and of his emotional and intellectual forthrightness.

Manabe's training was as a physicist, specifically a geophysicist, and all his intellectual sensibilities are those of a physicist. A late bloomer, especially by the standards of physics and math, he published his first major paper only at age 34. But over the next four decades, he would publish close to 150 more, almost always with one or another of his closest colleagues, and in time, those papers helped make a scientific revolution very nearly comparable to what happened, in the same period, in paleoclimatology.

Born in 1931 on the smallest of Japan's four major islands, Manabe grew up fairly sheltered from World War II, though he remembers seeing fleets of U.S. bombers high above, heading for the bigger cities. When he began studies at the University of Tokyo around 1950, he found a city still flattened, making for circumstances that were rather grim and austere even by graduate student standards. Working with just pencils, paper, and the old mechanical hand-cranked calculators, Manabe and a group of like-minded fellow students turned their attention to the solution of complex hydrodynamic equations, those that govern the behavior of fluids and gases. Such equations are key to understanding, among other things, the workings of the hydrogen bomb, and it was one of the H-bomb's fathers, ironically, who got the Japanese group's special attention and inspired much of its future work. In 1950, John von Neumann at Princeton's Institute for Advanced Study coauthored a seminal paper outlining the conditions for modeling and predicting weather or climate.[3] Von Neumann had been closely involved as a theorist in the design of the first major U.S. computer, the ENIAC, and as such was the closest U.S. counterpart to Alan Turing, the Englishman who had formulated during the war the theory of the programmable electronic computer. Von Neumann recognized that a computer would have to be able to store results, as calculations proceeded stepwise, and that insight was the basis of the ENIAC's architecture and that of all other computers to come.[4]

Like his close friend and colleague Edward Teller, von Neumann was a hard-line Cold Warrior, and he was interested in the possibilities of weather prediction and modification as instruments of war. But more than that, he considered weather—even more than bomb design—a problem that would stretch the capabilities of the newly invented electronic computers to their limits. He called weather forecasting "the most complex, interactive, and highly nonlinear problem that had ever been conceived of."[5] He put together a meteorology group at Princeton to address the challenge, and recruited a talented young man, Jule G. Charney, to head it up. Charney soon conceived a way of radically simplifying the basic hydrodynamic equations believed to govern weather and climate, so that they could be practically handled by the nascent electronic computer.[6] In 1950, Charney and colleagues created the first-ever computerized weather prediction, in a twenty-four-hour run on the ENIAC. They publicized the results in the journal *Tellus*, and their article got the attention of Manabe and his fellow students in Tokyo. In 1952, the U.S. Weather Bureau was persuaded to establish the Joint Numerical Weather Prediction Unit in Suitland, Maryland.[7]

Amazingly, several members of the alert Tokyo group soon migrated to the United States and, in short order, all played critical roles in helping make von Neumann's vision a reality. Manabe himself was recruited by a junior member of von Neumann's team, Joseph Smagorinsky, to work on the development of a climate model at a branch of the Weather Bureau in Washington, D.C.; the unit soon moved to Suitland, and finally got reconstituted as part of a larger organization as the General Fluid Dynamics Laboratory (GFDL) in Princeton. Another Tokyo graduate, Akio Arakawa, was hired by a pioneering climate theorist at the University of California–Los Angeles; there Arakawa developed mathematical techniques that—like Charney's initial breakthrough—proved indispensable in economizing on computer horsepower as models got much more complex.

Yet another of the Japanese students, Akira Kasahara, was lured to the National Center for Atmospheric Research in Boulder, Colorado, where he helped make NCAR as important a center for climate modeling as GFDL. Still another, Kikuro ("Kiku") Miyakoda, a longtime colleague of Manabe at GFDL, developed techniques that established the practicality of medium-range weather forecasting—predicting the weather more than just a few days ahead. Those ideas helped lay the foundation for the European Center for Medium-Range Weather Forecasting in Reading, England, as well as Reading's Hadley Climate Centre, which would emerge in the 1990s as one of the world's premier modeling centers. By then, Japanese science authorities evidently had noticed the unique contributions made by their compatriots, for they decided to have NEC build them the world's biggest supercomputer, which they dubbed the Earth Simulator. Briefly, they lured Manabe back to Japan to help administer climate modeling research, and briefly—for about five years—the Earth Simulator held sway as the world's most advanced supercomputer.[8]

What accounts for the signal Japanese contribution to the foundations of climate modeling? Pressed to speculate, Manabe is inclined to credit Japan's overpowering ethic of teamwork. "Everybody works extremely hard without selfish claims," and a "personal desire for glory can be accomplished only as a member of a team," as he puts it.[9] In Tokyo, each Japanese student would labor through the day and evening on his part of a problem, and at midnight they would assemble, compare results, and divvy up the next day's work. When high-performing members of that group were attracted to the United States by the prospects of better jobs, more generous funding, and—above all—ambitious scientific visions, they brought with them a group work ethic perfectly suited to high-performance computing and software engineering, which always is done by teams. But coming to the United States with its freewheeling individualism also was an enormously liberating and stimulating experience for them, Manabe remembers. No longer hemmed in by group pressures, positively encouraged to strike out on original paths, they were truly in a new world. Manabe himself was "a lone wolf," by his own estimation. Known among his fellow students in Tokyo as a rebel, he had occasionally gotten into trouble. For example, he referred to their professor's obsessive manipulation of hydrodynamic equations without regard for underlying physical realities as "numerical masturbation"—and unfortunately for Suki, the sarcastic observation somehow got back to the professor. Anyway, physical understanding of climate processes—not particularly math—was Manabe's forte. Mathematical wizardry he left to Arakawa, and he feels their skills proved complementary.

Though an individualist, Manabe did bring an acute understanding of collaborative work, and he is credited by all for having a shrewd understanding of who's best at what, which made him a superb coordinator and leader of their efforts. A hallmark of Manabe's career would be long and fruitful collaborations with a handful of especially close colleagues. With Richard T. Wetherald, he would systematically pursue hydrological issues, including soil moisture, cloud feedbacks, and carbon dioxide impacts. With Kirk Bryan Jr., he would investigate global interractions between atmospheric and ocean processes, especially the specific impacts of rising carbon dioxide, and the sensitivity of climate to perturbations. (On the basis of that work, Bryan would gain a reputation very nearly equal to Manabe's.) Somewhat later, Manabe also would develop fruitful working relations with Ronald J. Stouffer, with whom he would perform a series of increasingly refined "water hosing" experiments to see what happened after fresh water was injected into the North Atlantic; and with Anthony J. Broccoli, his fellow investigator of tropical cyclones, periodic dryness spells, the effects of topographical features like ice sheets and mountains, and the Milankovitch cycles. With a still younger colleague, Thomas R. Knutson, Manabe would evaluate tropical Pacific variability and the Southern Oscillation—a seesaw mechanism in the western Pacific that plays a key part in the genesis of El Niño and La Niña effects.

Throughout, Manabe has retained the loyalty and respect of colleagues, continuing to produce work of exceptional clarity and deceptive simplicity. Though he's quick to belittle his adopted scientific language as "broken English," and though he still has difficulty with English words containing awkwardly positioned *r*s, Manabe's writing in English has a lucidity and intelligibility often noted in other scientists of the first rank: for the nonspecialist and general reader, their writing is often easier to follow and appreciate than the work of lesser mortals.

In 1958, when Manabe joined the U.S. Weather Bureau's General Circulation Research Division at 615 Pennsylvania Avenue in Washington, his assignment was to help develop a realistic three-dimensional model of the atmosphere, in which changes over time would follow from the fundamental principles of physical motion and heat dynamics. Manabe's boss was a junior member of the von Neumann–Charney team, Smagorinsky, who had been put in charge of the climate model. Hiring Manabe, the science historian Spencer Weart has wryly observed, was perhaps Smagorinsky's "best idea."[10]

Manabe recognized that the first step toward a three-dimensional model would have to be a precise understanding of what happens physically in a single vertical column of air. With single-minded patience he set to work on the problem, which got him thinking about greenhouse gases. This was not because he was interested in global warming as such, but—quite to the contrary—simply because he realized that after solar radiation, the greenhouse gases had the greatest impact on global temperatures. Early on, Manabe discovered an error in the groundbreaking calculation by the Swede Svante Arrhenius, who had predicted in 1896 that doubling carbon dioxide would raise global mean temperatures by perhaps 5 degrees Celsius, and that cutting it in half would lower temperatures by a similar amount.[11] Manabe determined that Arrhenius and his intellectual descendants had neglected the crucial role of moist and dry convection in the atmosphere, that is to say, the vertical changes in temperature resulting from upward movement of warm air and water vapor.[12] Nevertheless, offsetting defects in Arrhenius's data concerning the way vapor absorbs infrared radiation had largely canceled his convection error, so that he had obtained fairly reasonable final results. Manabe's revised estimate of the amount surface temperatures would be boosted by a doubling of carbon dioxide was $2.3° C$[13]—roughly half Arrhenius's, but in the same ballpark.

With that result, the keystone for the construction of overarching three-dimensional climate models was in place. In 1965, Manabe and Smagorinsky, with colleagues, published two major papers, "Numerical Results from a Nine-Level General Circulation Model of the Atmosphere" and "Simulated Climatology of a General Circulation Model with a Hydrologic Cycle."[14] The model described in the papers, Manabe would later explain, treated temperature variation as a function not just of vertical radiative transfer and cumulus convection but also of large-scale horizontal air movements. Per Charney, it was driven by the basic equations of motion and thermodynamics, plus an equation for water vapor to account for both evaporation and condensation. The model also incorporated certain other macrophenomena, notably the reflectivity or "albedo" of large ice sheets. The model divided the atmosphere into a horizontal grid of intersecting lines 500 kilometers apart, and it radically simplified the earth's terrestrial and oceanic surfaces, dividing a dome-shaped world into just two big areas. The ocean surface was treated as a swamplike lake. Such "idealizations" allowed the model to be run in roughly one sixth the computing time that otherwise would have been required for a model with a comparable number of grid cells.

Using the model, Manabe and Smagorinsky came up with some basic estimates and predictions that have proved remarkably robust. They said that a doubling of carbon dioxide would lead to an increase in surface air temperatures of about $3° C$—exactly the midpoint of the latest projections from the Intergovernmental Panel on Climate Change (see box, page 118), which postulate a range of 1.5 to 4.5° C. They said that temperature increases would be significantly greater near the poles than at the equator—a trend that would be dramatically evident forty years later, with drastic thinning of Arctic ice. They said there would be an intensification of the hydrologic cycle—more water evaporating in the tropics, with more vapor being carried toward temperate regions and the poles. Those results have all stood the test of time, as models have become enormously more detailed and sophisticated and computational capabilities have increased many thousandfold.

Assessing these earliest achievements of Manabe and his closest collaborators, it's evident that besides breaking new ground in the study of climate, they were inventing a whole new way of doing science. Before, the model had been to test theory with observations, notes Jerry Mahlman, himself a leading climate scientist who worked with Manabe first as a junior colleague and eventually as his boss. "As the world began to respond to their pioneering work, the old paradigm...was replaced by the triad of observations, simulation, and theory."[15] That is, in fields like climate science, simulation came to rank with observations as a way of testing theory.

In hindsight, it's astonishing how well the almost ludicrously oversimplified Manabe–Smagorinsky model was able to simulate some tremendously important processes and trends. But to simulate climate dynamics more realistically, the next logical step was to treat the oceans more thoroughly, so that their role as a giant heat reservoir would be adequately recognized and their complex interactions with the atmosphere at least crudely charted. Working with his colleague Kirk Bryan Jr. in the late 1960s and early 1970s, Manabe developed what came to be known as a coupled ocean-atmospheric general circulation model (GCM), the template for all GCMs to come.

In 1969 they reported the first such model, lingering particularly on a little trick they had to use to take account of the radically different response times of the ocean and atmosphere to changes in heat. Pausing at the first stage of calculation, the effect of heat transfer in the ocean model was temporarily suppressed, but then, after a series of stepwise calculations, interactions of ocean and atmosphere were permitted. In this initial very simple model, which was otherwise essentially the same as the Manabe–Smagorinsky scheme, the oceanic and atmospheric parts were adjusted so that the evolution of the atmosphere in one year would be coupled with the ocean's evolution over 100 years. In later, more complex models, that ratio was widened even further, and new tricks were devised to prevent the models from generating absurdities—from running off the rails, so to speak—rather than arriving at specific equilibria for specific conditions.

Already by 1975, Manabe had a coupled model that he considered strong enough to test and evaluate climate changes resulting from increases in carbon dioxide. That year, he and his colleague Richard T. Wetherald published what has come to be seen as Manabe's second major work, "The Effects of Doubling the CO_2 Concentration on the Climate of a General Circulation Model." Again, it depended on some radical simplifications: little topography, no heat transport by ocean currents, and fixed cloudiness, among others. Even so, Manabe and Wetherald had enough confidence in their model to make detailed statements about the effects of doubled carbon dioxide levels on average temperatures, relative humidity, the water cycle, snow cover and albedo, heat balances, and kinetic energy carried by eddies in the atmosphere. They found that temperatures would rise in the troposphere (the lower atmosphere) but drop in the stratosphere. Surface temperature increases would be greater closer to the poles because the effects of more carbon dioxide would be magnified as reflective snow cover retreated, exposing more absorptive surface to the sun's radiation, and because a more stable atmosphere at the higher latitudes would better contain surface heat. They reaffirmed the earlier provisional conclusion that the hydrological cycle would be intensified.

Coincidentally, the paper appeared the same year that the first cumulative results came in from Mauna Loa in Hawaii, where Caltech's Charles D. Keeling had been monitoring the earth's actual carbon dioxide levels, at the behest of Revelle and Suess (see previous chapter). Looking back, Wetherald believes it was the combined effect of Keeling's data showing a steady rise of carbon dioxide since 1958 and the simulation reported in the Manabe–Wetherald paper that moved the subject of global warming into the scientific mainstream. Before, global warming had been widely seen by scientists as "more science fiction than science."[16] Now it was beginning to look like a compelling theory, if not yet established fact. Previously, if modelers had looked at warming at all, they had confined their attention to energy balances at the surface of the earth, which was an erroneous procedure and gave results that were all over the map. "We were the first to use an energetically consistent model," Wetherald says, setting the stage for the huge amount of research to come.

By 1979, global warming was being taken so seriously that the U.S. president's science adviser, a geophysicist, convened a panel of the National Academy of Sciences to report on the status of research. The panel was chaired by Jule Charney, who had set the ball rolling with his novel mathematical techniques in the early 1950s, and it focused largely on two models, Manabe's and an even simpler one developed by James Hansen (of whom there will be much more to say in the next chapter). Hansen, the lead scientist at the Goddard Institute for Space Studies, housed at Columbia University in New York City, had made up for unimpressive computational resources by using Arakawa's mathematical shortcuts to good effect.

Hansen's estimate of the change in the global mean temperature for doubled carbon dioxide was 4° C, while Manabe's most recent number was 2° C. The panel split the difference at 3 degrees, with a 50 percent margin of error—in other words, again, the exact range of 1.5 to 4.5° C ratified 2 decades later by the Intergovernmental Panel on Climate Change, with much greater statistical confidence. "We have tried but have been unable to find any overlooked or underestimated physical effect" that could reduce the estimated warming, the panel concluded.[17]

Some sense of what went into a supposedly simple Manabe model could be had from the Manabe–Bryan paper that provided the basis for the estimate incorporated in the Charney report:

> Velocity, temperature, water vapor and surface pressures are calculated on a global grid. Calculations are carried out at nine vertically spaced...levels.... For the computation of solar radiative flux, the seasonal variation of insolation is prescribed [that is to say, imposed by the modelers, rather than being internally generated] at the top of the model atmosphere. The depletion of solar radiation and the transfer of terrestrial radiation are computed taking into consideration clouds and gaseous absorption by water vapor, carbon dioxide and ozone.... The distribution of water vapor is determined [by accounting in time and in the spatial dimension for] advection of water vapor, condensation and evaporation. The process of moist convection is parameterized by the method of moist convective adjustment proposed by Manabe et al. [i.e., in his 1965 one-column model]. The ocean model is similar to the model of Bryan et al. [the one used for the 1975 doubled carbon dioxide experiment]. Velocity, temperature and salinity are calculated at each

grid point, and density is calculated from a realistic equation of state. In addition, the ocean model includes a simplified method of calculating the growth and movement of pack ice in polar latitudes.... 12 different levels are chosen so that they resolve the vertical structure of the ocean to a depth of 5000 meters.... For optimization of the computer time required to reach a state of quasi-equilibrium, 4.2 years time integration of the atmospheric part of the model is synchronized with 1200 years integration of the oceanic part.[18]

That was done by trading updated information between oceanic and atmospheric "data libraries," according to specified rules.

As models reached this level of complexity and subtlety, however much they were reviewed and endorsed by expert groups like the Charney panel, it obviously was essential to their credibility for their results to be testable against actual global observations. To be generally believed in the wider scientific community and ultimately by the general public, model outcomes had to correspond to the real world. As if to meet just that requirement, another landmark Manabe paper soon appeared, in 1981, coauthored by Manabe and his colleague Douglas G. Hahn: "Simulation of Atmospheric Variability." Relying on a variant of the usual coupled three-dimensional model but using newly devised mathematical techniques derived from "spherical harmonics,"[19] the simulation nicely reproduced the actual atmospheric pressures and their seasonal variations, by latitude, in the December–February period versus June–August. What was impressive about the paper, as a speaker noted during the January 2005 Manabe symposium, is that it "reproduced a large fraction of natural variability," and with quite unpredictable internal dynamics.[20] That is to say, before the model ran, it would have been hard to foresee how it would go about producing the results that corresponded so well to the real world, thus defusing possible criticism that Manabe and Hahn had somehow "cooked the books" to get the outcomes they wanted (see graph).

Perhaps just as significant was a 1979 paper by Manabe and colleagues reporting a good match of values generated by their model with actual observations of oceanic heat content. This was based on the first run of a coupled atmosphere-ocean model with realistic land elevations, recalls Sydney Levitus, another symposium speaker, and the ocean heat numbers were in startlingly good agreement with some 1.5 million oceanic temperature profiles that he himself had assembled.[21] These came from a variety of instruments suspended from research

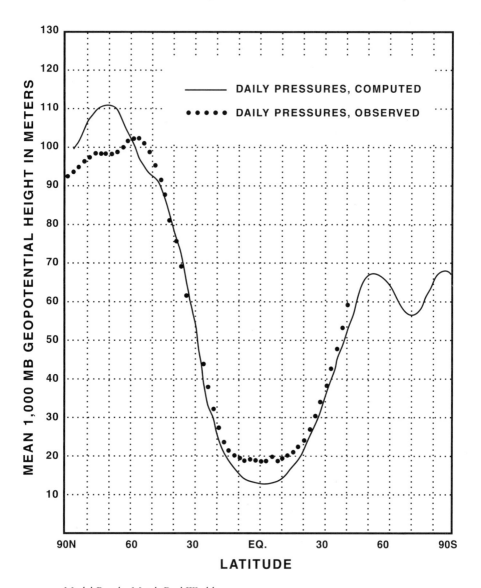

Model Results Match Real World

Source: Dansgaard, *Frozen Annals*, fig. 14.8

ships into the uppermost 250 meters of the oceans; temperature gradients were determined for each degree in longitude and latitude of the earth's surface; and the data were assembled at the National Oceanographic Data Center. Subsequent work, continued to the present day, has confirmed suspicions that the oceans are storing a significant fraction of the added heat trapped in recent decades by greenhouse gases.[22] Similar studies of ocean currents, which are driven by differences in temperatures and saltiness, have shown that most of the Pacific has been freshening in recent decades, everywhere except in the subtropical South Pacific and in the Indian Ocean. The Atlantic, in contrast, is freshening at deep levels toward the poles but getting saltier in shallow waters near the tropics.[23]

By the end of the 1970s, modeling work that had been done by just a handful of people in the whole world when Manabe and his fellow Japanese students migrated to the United States in the 1950s employed dozens of Ph.D. scientists, and by the end of the century would occupy hundreds. Improvements in the performance and availability of computers were just as impressive: by one measure, the computing horsepower available to GFDL scientists increased by a factor of three thousand from the late 1950s to the early 1980s, and in the next two decades it would increase again by at least as much. And that was just one laboratory.

Modeling centers with teams of specialized scientists were ramifying everywhere. In the United States, there emerged potent competitors to GFDL at the National Center for Atmospheric Research in Boulder, Colorado. By the turn of the century, it was equipped with the highest-performing supercomputer a U.S. manufacturer could offer, though its scientists eyed Japan's Earth Simulator jealously. Another important modeling group evolved at Germany's Max Planck Institute for Meteorology in Hamburg, and a still more significant one at the United Kingdom's highly prestigious Meteorological Office in Reading (the "Met Office"). There, in 1990, Prime Minister Margaret Thatcher established the Hadley Centre for Climate Research, which was soon recognized as one of the world's foremost scientific authorities in all matters pertaining to global warming. (Thatcher, perhaps because of her Oxford degree in chemistry, was quick to appreciate the new climate science and parted ways with her friend and political ally Ronald Reagan on the issue of global warming.[24])

With the proliferation of qualified personnel and high-performing supercomputers, it was possible now to conduct more ambitious computational experiments and seize more opportunities to compare results more or less in "real time" with real-world developments. In 1969, when Manabe and Bryan developed their first coupled model and ran the simulation on GFDL's machine for twenty-five full days, it was considered wildly extravagant. A decade or two later, a really sophisticated model might routinely run for up to six months or even a year, on enormously more capable electronic machinery, or simpler models might be run many times over, varying initial conditions and boundary parameters to see how much difference that would make to final results. By doing such "ensemble" runs, modelers could go a long way toward addressing issues associated with chaos: analysis of how individual outcomes deviate from the ensemble averages provides a measure of how much of the deviation is caused by changes in external variables, and how much is generated chaotically by purely internal dynamics. And by varying the values of factors that are uncertain, the scientists can also get a measure of a model's uncertainties.

People still mattered as much as or more than brute computational force. In 1978, Hansen's little group in New York City claimed it could accurately model the short-term cooling effects from sulfuric aerosols that had been released from the 1963 volcanic eruption of Bali's Mount Agung. Being a retrospective experiment—in which selective rearranging of past data might have been done to make the results come out right—the report understandably aroused interest but not necessarily belief. After the massive eruption in 1991 of Mount Pinatuba in the Philippines, however, Hansen boldly predicted just how much cooling could be expected globally in the next few years. By 1995 it was apparent that his forecast was dead on the mark.[25] (Those years following the eruption turned out to be virtually the only ones of the 1990s in which not much warming occurred. Otherwise, just about every year was the warmest ever recorded, making the decade as a whole almost certainly the hottest in at least a thousand years.)

The great general importance of the sulfate aerosols, from volcanoes and especially from coal combustion, was by now recognized. And although detailed understanding of their workings is one of the grand challenges facing modelers in the twenty-first century, at the most general level of analysis, one very important implication was pretty evident by the end of the twentieth. Aerosols from coal—together with heat absorbed by oceans—probably explained why the warming of the atmosphere in the last century or two has been only about half what the generally accepted models of greenhouse-gas impacts would have predicted.[26]

The sulfate aerosols also have been adduced to explain an anomaly that especially concerned scientists associated with the National Climatic Data Center in Asheville, North Carolina, who pointed out that in China, the former Soviet Union, and the United States, there have been increases in average nighttime surface temperatures in recent decades, but not in average maximum temperatures. Since the reflective sulfate aerosols only mitigate greenhouse warming effects during the day when the sun is shining, and do so the most during the summer months when the days are longest, their presence might keep down daytime and summer temperature maximums but leave minimums unaffected.[27]

Thomas Karl of the national data center and John Christy at the University of Alabama, Huntsville, also have drawn attention to a discrepancy between the earth's surface temperatures, which have been getting sharply warmer, and temperatures measured in the lower atmosphere, which have not. Those concerns, too, seem to be getting laid to rest. Researchers showed in 2004 that readings taken of the mid-troposphere by a satellite-borne instrument had been contaminated by temperatures from the lower stratosphere,[28] which cools with added carbon dioxide (as Manabe had shown in his early work). The instrument's unintended recording of lower stratospheric temperatures had led scientists to infer, incorrectly, that the troposphere was not warming at all. The issue was expected to be fully resolved in 2006, when a report is completed by a U.S. panel put together at the behest of the U.S. government to deliberately knock heads together.[29]

One of the most persistent trains of skeptical thought, propounded mainly by scientists outside the atmospheric modeling community, has been that solar cycles—not greenhouse gases—may be the main driver of the temperature changes seen in the last decades. But the most recent research suggests that the well-documented eleven-year solar cycle is too weak to have much effect on climate, and that a postulated longer-period cycle may not exist.[30]

Though it's hard to believe today, as late as the 1970s many scientists were concerned not about warming but, rather, the possibility that the earth might be trapped in a cooling cycle. The influential Russian climatologist Mikhail Budyko, also a fan of simple models and an important intellectual influence on Manabe, warned in 1968 that with just a small drop in temperature, insidious feedback effects from expanding polar ice sheets could produce another ice age or even lead to the world freezing over completely and permanently—"snowball earth," this nightmarish vision came to be called.[31] Other runaway scenarios, some formulated as early as the 1930s, anticipated catastrophic effects from diversion or even shutdown of ocean currents like the North Atlantic conveyor, which carries warm salty water from the tropics up to areas north of Iceland and southwest of Greenland, where the heavier saline water drops to the depths, to begin the reverse course.

To evaluate such possibilities, Manabe and his colleagues continued to churn out simulations, refining and sharpening estimates of how the earth would be affected by a doubling or even tripling or quadrupling of carbon dioxide, which seemed to them an increasingly likely prospect. They assessed some of the more credible ocean-current scenarios, and sought to test their models against data produced for the last ice age by ice drillers and ocean-floor boring teams. With Anthony J. Broccoli, for example, Manabe produced significant papers in 1984 and 1985 reporting on how well their models reproduced data obtained from CLIMAP, a global analysis of readings from ocean sediments sampled all over the world.[32] These experiments yielded, as one of the abstracts put it, "differences in ice surface temperatures and surface air temperatures which compare favorably with estimates of the actual differences in temperatures between the last glacial maximum [or LGM, 18,000 years B.P.] and the present." In their own way, the Manabe–Broccoli pictures charting their computer predictions against observations were as impressive as the matching of Dansgaard–Oeschger ice data against sedimentary observations. And though the CLIMAP data seemed at first to show unexpectedly little cooling in the tropics during the LGM, defying retrospective computer predictions, added observations made by Lonnie Thompson of Ohio State University in the Peruvian Andes seemed to resolve the issue in favor of the models.

Thompson, one of the last true adventurer-scientists, had taken a small crew with pack animals up a Peruvian mountain in the 1970s, where he discovered an exposed wall of the Quelcayya ice cap in which annual layers going back many hundreds of years were vividly apparent. (A photograph he brought back to show science funding officials in Washington still hangs, like an icon, outside many an office door at Ohio State's Byrd Polar Ice Research Center.) Thompson went on to produce, first from Quelcayya, then from many other glaciers, the single most important set of El Niño observations going back some 1,500 years. But when he first went to Peru, he intended to show how temperature trends at the tropics squared with those found by drillers working the poles—a project Dansgaard had dismissed as way too ambitious. In the end, Thompson got what he first was looking for and confirmed that tropical temperature trends were consistent with polar findings and model projections.[33]

Stimulated by suggestions from Wallace Broecker at Columbia University's Lamont-Doherty Earth Observatory, Manabe also turned his thoughts to possible consequences of a slowdown or shutdown in the North Atlantic thermohaline currents (huge currents that circumnavigate the globe, driven by heat convection and differences in saltiness at different sea levels and latitudes). He and Ron Stouffer, whose studies of this and closely related subjects are ongoing, found in 1995 that with a simulated "massive surface flux of fresh water to the north North Atlantic"—water hosing—"the thermohaline circulation weakens abruptly, intensifies and weakens again, followed by a gradual recovery, generating episodes that resemble the abrupt changes of the ocean-atmosphere system recorded in ice and deep sea cores."[34] Their results nicely simulated the Dansgaard–Oeschger and Heinrich–Bond cycles discussed in the previous chapter, producing large temperature changes in the immediate vicinity of the North Atlantic but not in the rest of the world, contrary to the more alarmist scenarios. (Scientists should be careful about how they use the word "global," cautions Manabe: while global impacts may be perceptible, this doesn't necessarily mean they're important.)

Wetherald and Manabe pushed on with increasingly precise simulations of how increasing carbon dioxide would affect the world regionally. As early as 1980, in a highly schematic model, they found that doubled or quadrupled carbon dioxide would have dramatically different impacts, leading to more river runoff at higher latitudes, a belt of decreasing soil moisture at mid-latitudes, and enhanced wetness along subtropical coasts. The differences between temperatures near the equator and those at the poles would narrow markedly, as ice and snow retreated and heat transported toward the poles increased substantially.

Those results were confirmed, but with considerable added specificity, in a more recent study by Wetherald, Stouffer, Manabe, and others.

"It has been suggested," that paper said, "that unless a major effort is made, the atmospheric concentration of carbon dioxide may rise above four times the pre-industrial level in a few centuries.... Our results suggest that water is going to be more plentiful in those regions of the world that are already 'water-rich.' However, water stresses will increase significantly in regions and seasons that are already relatively dry." With some understatement, they said, this "could pose a very challenging problem for water-resource management around the world." They anticipated much drier soil much of the year in regions that already are suffering protracted droughts, such as the southwestern areas of North America, northeast China, and the African grasslands. "In some of these regions, soil moisture values are reduced by a factor of two during the dry season. The drying in semi-arid regions is likely to induce the outward expansion of deserts to the surrounding regions."[35]

Surveying the art and science of climate modeling as it has evolved over the last half century, it would be hard to think of any individual other than Suki Manabe whose career has been so perfectly coterminous with the developing discipline, or who has contributed so much to it. NCAR's Jerry Mahlman, who worked for many years at GFDL with Manabe, first as a subordinate and later as the lab's director and leader of its stratospheric dynamics and chemistry program, said as much at a 1997 tribute, held to bid Manabe farewell as he left to take a managerial position in Japan's climate program. Without Manabe, Mahlman said, we'd be a lot further behind in climate modeling than we are. No other name, arguably, has come to be so intimately associated with the modelability and predictability of climate.

It's rather an irony, then, that Manabe is not altogether optimistic about the field's further advances in the next decades. This is not, or anyway not primarily, the attitude of an aging scientist reluctant to recognize that younger men and women may outdo him. When he contemplates the challenges ahead, Manabe speaks with sincere enthusiasm about how much "fun" it will be to take them on. He continues to take them on himself, exhibiting the childlike curiosity and enthusiasm characteristic of so many science greats.

So Manabe's skepticism about the near future of modeling has nothing to do with mental or physical exhaustion. Nor, as one might suppose, is it much related to the Lorenzian aspects of weather and climate—the chaotic element so often mentioned (usually, let it be said, by individuals who are not modeling specialists). Rather, Manabe's cautious pessimism is rooted in an acute awareness of how difficult good modeling is, and just how big the next challenges will be. One is treatment of clouds, and in his own carbon doubling and quadrupling experiments, he has found that outcomes can be rather drastically affected by just how clouds are handled. Human-made aerosols, which of course interact in important ways with clouds, are another. Though considerable progress has been made in understanding the action of sulfate aerosols, every anthropogenic aerosol added to a climate model brings a new element of uncertainty. The biggest complication of all may be the biosphere, which Manabe has always shied away from as way too big and complicated to model with confidence. Any number of times, Manabe has asked interlocutors how one could possibly hope to simulate all the plants and animals on earth in their interactions with the atmosphere and water, when just getting one raindrop right already presents formidable difficulties. As all those elements are incorporated into what goes by the name of "earth systems modeling," a development Manabe knows but just barely accepts is inevitable, uncertainties also will multiply. As a result, he believes, model outcomes are bound to get more unreliable and less credible "in the short run"—and that short run could turn out to be "forever," he warns.

In light of the startling discoveries made by the ice and ocean drillers, it would be reassuring if we could devise a credible model that simulated the waxing and waning of the ice ages believably, and at the same time projected what will happen with doubling, tripling, or quadrupling of greenhouse gas levels. But nobody is more aware than Manabe, who has handled many individual elements of that picture successfully, how far we are from a comprehensive model that can simulate both the known past and the unknown future. The inner dynamics of the ices ages remain shrouded in mystery, and any effort to simulate both them and present-day climate runs into computational limits. Even if we had a better handle on the specific physical chains of cause and effect that made ice sheets advance and retreat, even when supercomputers are much more capable than they are today, it still will not be possible to run the most sophisticated ocean-atmosphere GCM for the hundreds of thousands of years that would be required. And as earth system elements are added to models, that possibility may keep receding. Manabe notes that a Belgian and an Australian modeler once were able to simulate the ebb and flow of ice sheets by injecting known changes in carbon dioxide values into the model,[36] per Oeschger et al. But the results depended on some radical shortcuts and were not grounded in a good theory of actual physical relationships. He characterizes their model as "part heuristic"—as much a guide to further investigation as an attempt at realistic simulation.

Manabe is a man who feels pretty sure of what he knows but who also keenly feels the limits to knowledge. Both aspects of his intellectual character contribute to another irony—his skepticism about just how serious a problem global warming really is. Though Manabe was the first to treat the greenhouse effect just right, and though he is credited by lifelong colleagues with always having kept his eye sharply focused on the key roles of the greenhouse gases in climate, he does not exactly share their acute concern about where rising carbon dioxide levels are taking us. Though some regions will certainly suffer, as he and Wetherald have shown, he believes that others will gain; unless irreversible changes occur too fast for people to adjust adequately, it isn't obvious that the net effects will be negative. He thinks we may be underes-

When Manabe shows up at a top laboratory like the Lamont-Doherty Earth Observatory, it's an event. *Source:* Manabe family

timating our ability to adapt. In any case, he doesn't see how so many of his fellow scientists can make such strong statements about the dangers ahead.

Manabe remembers that once Budyko, the Russian meteorologist who anticipated the possibility of a snowball earth, came to the United States to accept an award that Manabe had helped him win. By then, Budyko had concluded that global warming was going to be a great boon to countries like Russia and Canada, so he seized upon the award ceremony as an occasion to sing warming's praises. Manabe found that a little amusing and a little embarrassing. But something like that was his reaction, too, at his own honorary symposium in San Diego, when his fellow modeler Hansen said that any warming greater than 1 degree Celsius would be "dangerous."

Building partly on Budyko's ideas, Manabe has convinced himself that the impacts of doubled or quadrupled carbon dioxide will quite definitely *not* be as severe as the consequences that followed from halving carbon dioxide levels. Manabe thinks of the processes that produced the ice ages roughly like this: the starting point would have been a large polar ice cover, like the one in the preindustrial Holocene; as temperatures dropped with falling greenhouse gas concentrations, the ice cover expanded around a very large periphery, so that the area of ice cover grew markedly, increasing the earth's albedo or reflectivity; that feedback lowered temperatures further, leading to additional spread of the ice around a still larger periphery, and so on.

But if one returns to that starting point and instead increases rather than decreases greenhouse gas levels, the effects of higher temperatures are quite different. The ice cover contracts, but for a given change in the ice cap's radius (from pole to periphery), the change in surface area with contraction obviously will be smaller than the change with expansion. And stepwise, with each increment of temperature and gas increase, the change in ice surface will get less significant. Accordingly, the warming feedback from polar ice changing to water will be weaker than the cooling feedback from water changing to ice.

To take the extreme cases, if greenhouse gases and temperatures drop enough, the final outcome will be Budyko's lifeless snowball earth. But if they go up enough, all ice may disappear from the world and a lot of other things may change too, but life will still be possible.

This is a compelling train of thought but also, as Manabe would be the first to concede, one that leaves a lot out. So it's perhaps no surprise that many who have closely followed his work over the years, and even many of those directly involved in it, do not necessarily reach the same final conclusion. One such person is Mahlman, now a senior research fellow at NCAR in Boulder. "If the climate model projections on the level of warming are right," he told *The New York Times* in an interview, "sea level will be rising for the next thousand years, the glaciers will be melting faster, and dramatic increases in the intensity in rainfall rates and hurricanes are expected. It means a summer drying out of the interiors of continents, with a threat to agriculture systems, planetwide. In the winter, it will rain more in our [U.S.] latitudes.... If sea levels rise as fast as we think they will, the Florida Everglades are doomed. Low-lying countries like Bangladesh and Holland will be in serious trouble. And you can say goodbye to any island that were formed by coral."[37]

As for the Kyoto Protocol, which has been rejected by the United States as too onerous and too unfair, it's no more than "a valid first step," as Mahlman sees it. "Thirty Kyotos might do the job," he says.

CHAPTER 7

The Synthesizers

A S DRILLING AND MODELING have evolved from tiny pioneering efforts into global enterprises, the scientific results have become so voluminous and complex that it would take a Renaissance man or woman to digest them as they are reported. A handful of individuals may be capable of staying on top of it all, but even they can be stretched beyond their limits when it comes to drawing practical conclusions and communicating them to policy makers and the public.

Because of this, the trend in recent years has been for universities to found a new kind of scientific institution, dedicated to synthesizing all scientific knowledge bearing on the earth's history and destiny and to putting that knowledge into a form fit for general human consumption. Inevitably, though, such outfits are dependent, especially at their inception, on their scientific members who are best at integrating vast arrays of information from disparate sources. Inevitably, too, patrons influence the direction of research, and even the conclusions reached.

One fledgling institution, the Global Climate and Energy Project, was founded in late 2002 at Stanford University, in Palo Alto, California, with funding from Exxon Mobil, General Electric, a large German energy company called EON, and the oil drilling company Schlumberger.[1] The four companies said they were giving Stanford $225 million over 10 years to do research on how to meet the world's energy needs without worsening global warming. Exxon Mobil, which already had spent millions of dollars to propagate the view that climate science is too uncertain to demand costly changes in energy policy, pledged $100 million.

Stanford boasts some of the biggest guns in climate science, people like the bioclimatologist Stephen Schneider, who sometimes illustrates lectures with a cartoon portraying the price of a gallon of gasoline as equivalent to the cost of the U.S. aircraft carrier guaranteeing its safe delivery. Nobody would accuse a person like Schneider of allowing his work to be influenced by oil-company donors. But one may still wonder whether the Stanford project will always be able to resist the influence of its corporate benefactors.

At the time the Stanford grant was announced, an institute with similar goals but a somewhat different political complexion was taking shape on the eastern seaboard, at Columbia University in New York City. There, the notion had germinated in the mid-1990s of creating an "earth institute" and promoting it as a magnet for new talent and a lodestone for policy makers seeking to guide the world toward safer harbors. After a shaky beginning, in mid-2002, Columbia lured a star economist and celebrity policy adviser, Jeffrey Sachs,[2] from Harvard University to take over the Earth Institute's leadership. Basically, his job was to leverage Columbia's scholarly expertise in geology, the environmental sciences and engineering, and climate science into something that would be much bigger than the sum of its parts.[3] If Sachs wanted to expand the institute's scope to embrace general economic policy, particularly the economics of public health in poor countries—his main personal interest in recent years—that would be all right with Columbia. But a main focus, it was clear, would be similar to Stanford's—figuring out how to power the world's economies for the remainder of the twenty-first century without destroying the earth's environment in the process.

When the Earth Institute was founded in 1996, Columbia had two big assets it wanted to capitalize on: the nation's oldest school of mines, which had evolved into a general program of environmental engineering (how to obtain, process, and dispose of all kinds of materials properly); and what used to be known as the Lamont-Doherty Geological Laboratory, now the Lamont-Doherty Earth Observatory, up the Hudson River north of New York City, in Palisades.[4] Though Lamont-Doherty had originated as a home base for scientists who spent most of their real time out in the field pounding rocks with little hammers, it had been much more than an old-fashioned geology department for many decades. If, for example, you wanted to know whether seismologists would be able to adequately monitor a comprehensive world ban on nuclear weapons tests, then the person to talk to was Lamont-Doherty's Lynn Sykes, perhaps the world's top seismologist. If you wanted to know how the world's oceans and atmosphere might have interacted during the ice ages, sending glaciers all the way south to what is now Cincinnati and burying what is now Chicago under hundreds of meters of ice, you asked Wallace Broecker, one of the world's leading experts on paleoclimatology.

When Sachs arrived on the scene in mid-2002, his first order of business was to raise about $200 million to support the Earth Institute's planned research program—roughly matching, coincidentally or not, the funds pledged to Stanford by its four big corporate donors.[5] An even bigger but equally challenging problem facing Sachs was to figure out how to get the institute to speak with a single voice on at least some of the key issues it was claiming as its own. For the Earth Institute's explicit mission, to borrow the famous programmatic statement of Marx and Engels, was not merely to understand the world, but to change it—and change it for the better. Having first achieved international celebrity around 1990 as the most influential advocate of "shock therapy" financial policies in South America and Eastern Europe, where national economies were crippled by staggering inflation rates, Sachs was seen as a man unafraid to ruthlessly apply theory to practice. And having turned his attention in the late 1990s to the chronic and debilitating diseases afflicting the poorest countries of Africa and Asia, making himself a forceful advocate for large-scale, well-focused international aid, Sachs again had shown himself to be an effective spokesperson for the ideas he believed in. But how could even Sachs get an institution to take a clear, unified position on the hugely controversial question of global climate change, when that institution was almost as divided on the issue, in some ways, as the public at large?

Putting oneself in the shoes of Sachs, and surveying the Earth Institute roughly as it would have looked to him when he moved to New York in mid-2002 to take charge of it, at one extreme was Lamont-Doherty's Wally Broecker, the far-ranging geochemist who generally is credited with having given the Dansgaard–Oeschger cycles their name. Broecker sometimes seems to take such an alarmist position about global warming, one wonders whether he thinks anything really can be done about it. The solutions he visualizes have about them an air of desperation.

At the opposite extreme is Jim Hansen, director of NASA's Goddard Institute for Space Studies (GISS), which is housed at Columbia and is affiliated with the Earth Institute. Hansen argues that the Kyoto Protocol measures could be quite helpful and not too costly, and that progress made to date in reducing greenhouse gas emissions has been much more significant than generally appreciated. He has little use for those, like Jerry Mahlman, who say it would take "thirty Kyotos" to make any real dent in the problem of climate change.

Broecker and Hansen are but two of many world-renowned climate scientists at the Earth Institute, but it seems fair to say that they are the poles that pretty well define everybody else's orientation. Both are unusual individuals, with remarkable achievements and a legitimate claim on any serious student's attention. Hansen is the more modest of the two, and yet has never been shy about taking what he has to say to the general educated public, which he did most recently in the article "Defusing the Global Warming Time Bomb" in the March 2004 issue of *Scientific American*. The much more bombastic Broecker, ironically, is less fond of publicity, and yet in 2004 his influence was evident in a leaked Pentagon intelligence report warning that drastic climate change could occur very quickly and quite soon, leading to widespread flooding, drought, famine, and war. "Nations with the resources to do so may build virtual fortresses around their countries, preserving resources for themselves. Less fortunate nations...may initiate struggles for access to food, clean water, or energy," the report warned. (Because of such language, Broecker repudiated the report after it started to circulate.)

Who are Hansen and Broecker, and how have their views come to have such influence? Hansen[6] was born on a farm in Iowa and grew up in the small town of Dennison, where he was the high school science star. He enrolled for undergraduate and graduate work in astronomy at the University of Iowa, where he completed his Ph.D. under the supervision of James Van Allen, a scientist quite famous at the time for having discovered the Van Allen radiation belts circling Earth. Hansen's field of expertise was planetary atmospheres, specifically the Venus atmosphere. When he joined Columbia's Goddard Institute for Space Studies in 1967, it was to continue his work on Venus's atmosphere, which he came to think of as a runaway greenhouse—so severely heated by radiation-trapping gases, it would never cool to levels required to sustain life.

In the 1970s, Hansen was principal investigator for an important experiment that would head for Venus aboard a NASA space probe. Before that mission had arrived, however, his interest had shifted to another planet's atmosphere: the Earth's. He resigned as leader of the Venus experiment to devote himself, initially, to studying the role trace gases play in Earth's atmosphere, and how the changing composition of that atmosphere might affect Earth itself. This endeavor would soon make him, like his mentor Van Allen, famous. A completely unassuming man, Hansen today occupies the top-floor corner office at GISS's headquarters on upper Broadway in Manhattan—coincidentally, directly above the real Tom's Restaurant, the fictional television setting for many a self-absorbed conversation among Jerry Seinfeld, George, Kramer, and Elaine.

What brought Hansen to national prominence was testimony he gave to Congress in summer 1988, in which he warned that human activity was warming the earth, possibly dangerously. Even though at this time the "greenhouse effect" was well understood and universally accepted, in the sense that everybody knew that without an atmosphere trapping radiation, earth would be much colder, most scientists were very reluctant to claim that gases like carbon dioxide and methane—mainly emitted as by-products of combustion—were making the world significantly warmer. In fact, scientists had only just begun to recognize and acknowledge that the world was warming at all, and to the extent they did, it was in no small part because of Hansen's research—and Broecker's as well. (In 1975, Broecker presciently wrote that the cooling trend still prevailing might soon, "within a decade or so, give way to a pronounced warming induced by carbon dioxide."[7])

In 1981, Hansen and colleagues published papers taking issue with the conventional wisdom that the earth was destined to keep cooling. In one, "The Climate Impact of Increasing Carbon Dioxide," published by the preeminent U.S. research magazine *Science*, Hansen showed that the world had become about four tenths of a degree warmer in the previous century. In another paper, Hansen observed that average sea levels had risen about 12 centimeters in the past hundred years, and that not just melting ice but thermal expansion of the water itself was an important factor. Evidently he was one of the first to notice what is now considered a rudimentary element in the warming picture.[8]

The 1981 papers, especially the one in *Science*—which was rejected twice by that journal and once by Britain's *Nature*, before *Science* finally agreed to publish what it had considered a much too long report—got wide attention. Both *The New York Times* and *The Washington Post* ran editorials inspired by it, and environmental activists based in Washington, D.C., like Rafe Pomerance of Friends of the Earth, began to talk Hansen up. He was introduced to Al Gore, then a Tennessee congressman taking a serious interest in environmental issues, and invitations to testify before congressional committees followed. When he appeared before committees, unlike so many of his overly cautious colleagues in the field of atmospheric science, Hansen was willing to go where almost all others still feared to tread. Yet even then, a prudent and careful man, he walked a fine line.

Hansen told a Senate committee in 1988 that the earth was almost certainly getting warmer and that energy use was almost certainly the main cause. But when senators pressed him with questions about whether, for example, parched conditions in the American Midwest were caused by global warming, he confined himself to saying that while climate models showed droughts were becoming more likely, the occurrence of any specific drought still was a matter of chance. As he refined that statement in subsequent testimony, he compared the odds of there being a drought to throwing dice, with one set of outcomes representing drought, another set no drought. It was as though, he explained, we were loading the dice in advance to make the drought outcome more likely each time we threw them.[9]

Hansen now has stepped forward again into the limelight. But today he is arguing that the climate problem, though very serious, is not as intractable as many of his fellow scientists would have us believe. And though many of his peers might prefer to ignore him, because they worry that his new position will encourage complacency and support policies of inaction, they know they cannot dismiss him. In the fall of 2001, he and colleagues put forth what they called an "alternative scenario," which looks roughly like this. Hansen et al. start with the generally shared observation that increases in the main greenhouse gases—carbon dioxide, methane, the chlorofluorocarbons, and nitrous oxides—warmed earth by roughly an additional 2.5 watts per square meter between 1850 and 2000. That heating was partially offset by aerosols—particles suspended in the atmosphere, such as sulfates from combustion, added cloud cover, and volcanic dust—which reflect sunlight, preventing its energy from being trapped in the atmosphere. The net effect, between 1850 and 2000, was an increase of 1 watt per square meter on earth and a global temperature about 1 degree Celsius warmer than it otherwise would have been.

To visualize that more concretely, it is as if we divided up the entire surface of the globe into squares, each 10 by 10 meters, then lit a 100-watt lightbulb in each of those squares and burned the bulbs around the clock. (Incandescent bulbs convert almost all their energy to heat—that is to say, they waste it—and only a tiny fraction to their intended light.) In other words, each bulb would give off 100 watts of heat per 100 square meters, or 1 w/m^2. It's easy to sense that over time, this buildup of heat would have a noticeable effect on the whole earth's temperature.

Looking ahead, the Intergovernmental Panel on Climate Change (IPCC) has estimated that unless we make concerted efforts to change our ways, we will add another two to four watts of heat per square meter of earth in this century, boosting the globe's temperature—as the impacts of all the accumulated gases make themselves felt—by as much as one degree per watt. This would be like burning as many as four additional 100-watt bulbs around the clock in each of the earth's imaginary 10-square-meter compartments. (A UN-sponsored organization that has involved hundreds of climate scientists over the past two decades, the IPCC was set up to produce consensus estimates of what can be expected from greenhouse gas emissions. See "What Is the IPCC?")

Though Hansen supports the IPCC and the general thrust of its work, he disagrees with its conclusions in some important respects. He feels the IPCC scientists have relied too much on computer projections and too little on actual measurements of recent trends, which have been more encouraging in some ways than might be supposed. And he thinks that IPCC reports have underestimated the net impact of aerosols, specifically that of heat-absorbing "black carbon," basically plain old soot. He thinks that black carbon has contributed much more to warming than is generally appreciated and that its role has been counterbalanced (and hidden) by other reflective aerosols, notably the sulfates related to acid rain. The conclusion he draws for policy is that it would be better to focus specifically on reducing soot and certain greenhouse gases such as methane and ozone than to concentrate almost exclusively on reducing the main greenhouse gas, carbon dioxide. This is because he thinks reducing carbon dioxide adequately would require heroic cuts in fossil fuel consumption, which the world so far has been demonstrably unwilling to make. "If sources of CH_4 [methane] and O_3 [ozone] precursors are reduced in the future...[and that were] combined with a reduction of black carbon emissions and plausible success in slowing CO_2 emission, this...could lead to a decline in the rate of global warming, reducing the danger of dramatic climate change," Hansen and his collaborators wrote in their 2000 paper describing their alternative scenario, which the U.S. National Academy of Sciences published (see figure).[10]

Hansen's alternative scenario unleashed a storm of criticism, mainly because it struck some fellow experts as a brief for complacency, since it suggests heroic cuts in carbon dioxide emissions may not be necessary after all. But that reaction surely was unfair. In a version of the alternative scenario that Hansen presented to the president's Council

---

What Is the IPCC?

The Intergovernmental Panel on Climate Change was established in 1988 by the World Meteorological Organization and the United Nations Environmental Programme to assess on a scientific basis all the best evidence having to do with human-induced climate change. It does not conduct research as such, but rather draws upon peer-reviewed scientific studies to reach its conclusions. According to the IPCC, its "reports seek to ensure a balanced reporting of existing viewpoints and to be policy relevant but not policy prescriptive."

Each of the panel's reports is subjected to the critical scrutiny of scientists all over the world; when the reports pass muster with them, they are subjected to a second round of scrutiny by government experts. The IPCC's work is divided among three groups, headed by independent scientists and scholars, to assess the science of climate change, the likely effects of warming, and options for limiting greenhouse gas emissions. The IPCC also has a task force that gets countries to develop complete inventories of their greenhouse gas emissions. (The U.S. inventory provided the foundation for much of the analysis in chapter 3 of this book.)

The first comprehensive IPCC report appeared in 1990 and was the foundation for the United Nations Framework Convention on Climate Change, adopted at Rio de Janeiro in 1990. The immensely influential second report, in which "a human fingerprint" was first identified in climate change, led to the adoption of the Kyoto Protocol, to implement the convention, in 1997. The third report, which was completed in 2001, represents the best summary of knowledge about climate change available at present. The fourth will appear in 2007.

Information about the IPCC and pdf files of most of its important literature are available at www.ipcc.ch.

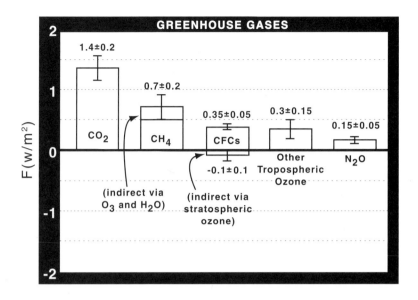

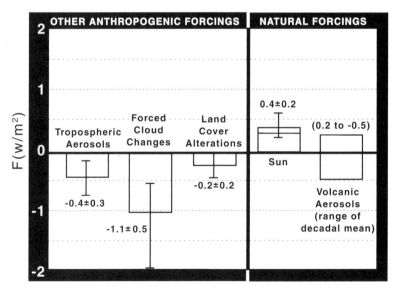

Estimated Climate Forcings Between 1850 and 2000

Source: Hansen, Sato, Ruedy, Lacis, and Oinas, *PNAS* (2000)

on Environmental Quality in mid-2003, he made clear that his intention was quite the opposite: he considers the IPCC's projected warming of 2 to 5 degrees Celsius to be intolerably large, as catastrophic rises in sea levels cannot be excluded in that temperature range; the whole point of his scenario, accordingly, is to keep net added forcings (climatologists' term for changes that cause other changes) below 1 watt per square meter in the next 50 years.[11] To some, since Hansen's scenario emphasizes cutting gases other than carbon dioxide, it may even have looked like an open invitation to just keep burning more fossil fuels. But that too is unfair. In a 1990 paper, "Sun and Dust Versus Greenhouse Gases," published in Britain's *Nature* (probably the world's most prestigious scientific magazine), Hansen addressed the question of whether, since sulfate aerosols resulting primarily from coal combustion counteract and mask the warming effects of the greenhouse gases, we should just burn much more coal. Along with almost all others who have considered the question, he concluded that this would be a deal with the devil because it would work only as long as we kept burning more and more coal; as soon as we stop, the aerosols would wash out of the atmosphere, leaving all the long-lived carbon dioxide that had accumulated from coal combustion to suddenly make its influence felt.[12] The result could be an abrupt, devastating increase in global temperatures.

In the latest revised versions of the alternative scenario, Hansen has stressed that cuts in black carbon emissions would only compensate, roughly, for cuts in sulfate aerosols expected to result from clean air regulation—that is, the absorptive effects of black carbon roughly equal the reflective effects of the sulfate aerosols. Still, even the reformulated scenario relies on some rather optimistic assumptions about what can be realistically accomplished in the next 50 years. A closer look at where the soot comes from reveals that half actually is from biomass combustion: forest fires, both natural and deliberate; and widely dispersed burning of crops in places like China, India, and Africa, where the practice often is crucial to rural economies, for both energy and soil replenishment. However undesirable such practices may be—including Amazonian deforestation, done in the name of development and prosperity—they are not easily reined in. In the industrial countries, diesel vehicles are a major source of black carbon, highlighting a nasty trade-off: since diesel engines are more fuel-efficient than gasoline-powered ones, the general trend in some parts of the world is to rely on them more—despite their detrimental effects on air quality—both to save energy and to cut carbon dioxide emissions. What is more, even the half of the black carbon coming from industrial processes in industrial countries is not mainly from power plants. So getting rid of soot is not just a matter of attaching scrubbers to electricity generators in rich countries.

Hansen's optimism about reducing methane emissions by up to 30 percent runs up against similar quandaries. Most human-related methane emissions are associated with very widely dispersed farm activities—above all, rice cultivation and livestock breeding. How quickly can existing practices be modified all over the world? As for the emissions associated with natural gas leakage, as the world relies ever more intensely on this fuel, will it be possible to simultaneously upgrade enormously dispersed distribution systems?

If Hansen's competence were in dispute, or if he were the kind of person who habitually seeks the spotlight and is willing to say anything to get into it, his views might have been dismissed by his colleagues. But his abilities have never been in question—even as a high school student in Iowa, a fellow pupil has attested, his problem sets in math and science courses were so exemplary they were used as models for the whole school; he ended up virtually teaching some of the classes. As for his character, he has been described by a fellow atmospheric scientist as being "almost entirely without ego," as almost "unique among scientists in seeming to genuinely not care what anybody thinks of him."[13] Hansen is the very opposite, in a word, of those famously self-absorbed Seinfeld characters with whom he shares Tom's Restaurant downstairs from GISS.

To this day, Hansen's group continues to churn out authoritative estimates of earth's energy balances, based on the streamlined modeling style that has been their hallmark for decades, and to draw important and sweeping conclusions from those estimates. In one of its more recent reports, published in the April 28, 2004, issue of *Science* magazine's online "*Science Express*," Hansen and his collaborators found that because of delayed impacts from greenhouse gases already in the atmosphere, about 1 degree Fahrenheit or $0.6°$ C of warming is "in the pipeline"—destined to take place in the decades ahead. "This delayed response," they wrote, "provides an opportunity to reduce the magnitude of anthropogenic climate change before it is fully realized, if appropriate action is taken. On the other hand, if we wait for more overwhelming empirical evidence of climate change before acting, the inertia implies that still great climate change will be in store, which may be difficult or impossible to avoid."[14]

But Hansen's is not the only view of the warming problem and his preferred solution not the only proposed means of making a radical attack on it, without having to give up our carbon dioxide habit. Wallace Broecker, the Earth Institute's other superstar climate scientist, takes an essentially similar view of what global warming portends but draws much more drastic conclusions for policy. Like Hansen, Broecker hails from the American Midwest. He has the same casual personal style, rarely if ever donning a tie or jacket. Once, when he was photographed for a feature article that a leading scholarly journal was running about celebrity geoscientists, he was the only one, he later said, to appear for the shoot in jeans and an old sweater, his habitual attire. Like Hansen, he leads a simple personal life, and does not much care for putting himself in situations where he would be at the mercy of ambiguity or subjectivity—the arbitrariness of mere opinion or personal whim. Yet Broecker openly appreciates his own achievements and worth, and he considers himself well entitled sometimes to take extreme views. Whereas Hansen is telegenically calm and cool, Broecker often comes across as impassioned, even intemperate. When he's angry, in fact, it's not always clear what's really madness and what's method. When he was still a graduate student, he has said,[15] he had occasion to meet Hans Suess, a pioneer in radiocarbon dating—and as it happens, one of the more competent veterans of Nazi Germany's wartime nuclear weapons effort (see box, "The Real Dr. Suess"). Suess, evidently recognizing at a glance Broecker's precocious talents, gave him some advice that he evidently took to heart: at all costs, avoid administrative responsibilities. The best way to do that, said Suess, would be to just do something that impressed his peers as really crazy a couple of times a year. That way he would never be bothered with committee or departmental chairmanships and would be able to single-mindedly pursue his interests.

Broecker followed that advice to the letter, dedicating his whole career to pure research, so that when he reached age 69 in the year 2000, he could boast of 385 publications in his previous 45 years, or 9 per year. With a characteristic mixture of self-effacement and self-aggrandizement, he illustrated that productivity in an autobiographical sketch by constructing a bar graph in which each year's scholarly output was repre-

The Real Dr. Suess

During World War II the German-born Hans Suess was a member of a three-man team in Hamburg that developed more efficient ways of producing heavy water (water in which the usual hydrogen atom is replaced by a deuterium atom) and methods of enriching uranium by means of centrifuges. Heavy water, which at that time was produced commercially only in Norway, could be used in a nuclear reactor instead of graphite to produce plutonium for a bomb; alternatively, uranium could be enriched to provide the fissile material for the bomb. Thus, the Hamburg group played a critical role in Nazi Germany's effort to determine whether the country could build an atomic bomb in time to influence the outcome of the war.

The Hamburg team, which was led by Paul Harteck, stood out for its shrewd common sense and its technical competence in a program otherwise plagued by strategic blunders and fundamental misconceptions. As Mark Walker, the most evenhanded and reliable of the historians who have studied the program, put it, "With the support of his younger colleagues Wilhelm Groth and Hans Suess, Harteck was always ready with an answer to a new problem" (*German National Socialism and the Quest for Nuclear Power* [Cambridge: Cambridge University Press, 1989], 146). Progress "was largely due to the impressive skill, energy, and devotion of Paul Harteck and his assistants Wilhelm Groth and Hans Suess" (149).

After the war Suess immigrated to the United States and, at the Scripps Institution of Oceanography in 1957–58, played a key part in initiating the direct measurement of carbon dioxide levels in the atmosphere (see chapter 4). He also identified a phenomenon, now known as the Suess effect, that has played a role in determining preindustrial carbon levels. It occurred to Suess that, since all the carbon 14 contained in fossil fuels is fully decayed (having been there for millions of years), when carbon dioxide is emitted from coal, oil, or gas combustion, the effect is to dilute the concentration of carbon 14 found in the atmosphere.

The numbers are rather startling: about 7.5 kilograms of new carbon 14 are produced annually by cosmic radiation; at present, by comparison, about 5 quadrillion (5 to the 15th power) kilograms of fossil carbon are pumped annually into the atmosphere.

sented by a Number 2 pencil—the advanced tool he had used to write all his papers. It is worth lingering over the fact that Broecker has never composed his papers on a typewriter or a PC. He in fact doesn't much care for the computer even as an instrument of scientific calculation. He prefers to look directly at the physical realities that interest him, doing his calculations back-of-the-envelope style. Partly because of that scientific approach, perhaps, he is able to vividly convey what he thinks he has learned.

Broecker was born into a family of modest means in the Chicago suburb of Oak Park, Ernest Hemingway's hometown. He enrolled at Wheaton College, a respectable local institution of higher learning, but certainly no Caltech or MIT. Midway through, a fellow student with similar interests suggested he apply for a summer internship at the Lamont-Doherty Geological Laboratory. Broecker headed off to the Palisades, and he arrived for his first day of work at the laboratory on June 15, 1952. He would never leave again, except to do field work, which he would do everywhere.

Almost immediately, Broecker was initiated into the new art of carbon 14 dating, the hugely powerful tool that enabled scientists to effectively address problems that had interested but defied them for decades or centuries—how the oceans' waters mix over time, when coral reefs originated and how they grow and die, when prehistoric populations arrived at the places where their bones were found. Broecker soon proved himself a grand master of the technique, and during the coming decades, he would follow it wherever it took him, becoming in the process a recognized expert not only on the dating method itself but also in a remarkable range of otherwise unrelated fields. When he started in 1952, the number of people specializing in the kinds of measurement he did numbered no more than a few dozen; by 1982, when he published an overview of one aspect of his work, *Tracers in the Sea*, the field had grown to encompass thousands or tens of thousands of scientists working just about everywhere in the world.

At the time Broecker published *Tracers in the Sea* (which he did in the literal sense, shunning the services of a commercial publisher and founding a company to print the book himself), he was fighting third-stage lymphoma. Had he died, he would be remembered as a leading geochemist—a top specialist, but a specialist just the same, known really only to his scientific peers. But the work he produced in the next decade thrust him to the forefront of researchers studying the history of the world's climate, so that by the time Sachs arrived at the Earth Institute, some of Broecker's peers had begun to think of him as a near genius. For by now, he had established himself as an individual with a singular talent for synthesizing and integrating findings from all over the map, and for arriving at startling and unsettling conclusions.

By the mid-1980s, Broecker was giving lectures to general audiences in which he lined up all the best data obtained in the previous decades from chemical and radiochemical analysis of tree rings, coral shells, pack rat middens, ocean sediments, and ice cores. He'd show that the data, obtained by thousands of scientists working independently in many parts of the world, were mutually consistent and reinforcing. He'd show that when measures of temperature and atmospheric competition from ocean floor bore holes and ice cores were plotted together on the same graphs, it was clear that levels of carbon dioxide and methane and temperature increases had moved in lockstep for the last 430,000 years. Soon, graphs like Broecker's and information about his concerns began appearing in newspaper op-ed columns and books, like former Senator Al Gore's *Earth in the Balance* (1988). Meanwhile, back at Lamont-Doherty, Broecker himself was brooding over why, given relatively slight changes in radiative gas forcings, the earth has been subject to such cataclysmic climate events. Broecker suspected that the oceans, especially sensitive to temperature, might be the missing link. Specifically, he thought the culprit might be changes in the global circulation of water around the world, a phenomenon he had first studied as a young researcher at Lamont-Doherty, when he was asked to use radiocarbon dating techniques to figure out how fast deep oceanic waters are replaced by water mixing in from the surface. That work led to the charting of what is now known as the thermohaline circulation (the huge currents that circumnavigate the globe, driven by heat convection and differences in saltiness at different sea levels and latitudes). One big part of that is the great North Atlantic conveyer, of which the Gulf Stream is but an appendage. In effect, as the saltier and denser cooling waters in the North Atlantic sink, warm surface waters are drawn north, warming Northern Europe.

Everything came together for Broecker around 1985, when a fellow scientist suggested that the world's climate system appeared to be oscillating between two radically different states or modes of operation. Drawing on his studies in oceanography and paleoclimatology, Broecker hypothesized that the change might be due to the sudden turn-off of deep water formation in the North Atlantic. In 1987 he published two seminal papers. "Unpleasant Surprises from the Greenhouse," a scientific article published by *Nature*, described how, by adding large amounts of greenhouse gases to the atmosphere, for all practical purposes using it as an unregulated sewer, we might trigger yet another abrupt climate change. "The Biggest Chill" was a more popular article that appeared in the American Museum of Natural History's magazine, *Natural History.* "I outlined in layman's terms how [the last and most recent of the earth's cataclysmic climate events] appeared to have been triggered by the sudden release into the Labrador Sea of a huge amount of ponded glacial melt-water," Broecker later recalled. "Such an injection would have reduced the salinity of the water in the region where deep water forms to the point where they were no longer dense enough to sink to the abyss" and draw warmer waters north.[16]

The story Broecker told in "The Biggest Chill" took place in the period known as the Younger Dryas, about 12,500 years ago (after a Scandinavian flower that shifted its habitat then) and involved a huge, frozen inland lake (Lake Agassiz, named after the nineteenth-century Harvard geologist who pioneered the study of the ice ages), which stretched from what's now Minnesota and the Dakotas well into Canada. Lake Agassiz suddenly melted, sending a flood of fresh water through what's now the St. Lawrence Seaway into the North Atlantic, making its water too salt-free to properly sink. The great North Atlantic conveyor shut down. The last bitterly cold spell ensued.

Broecker spelled out the implications of this kind of scenario in his other 1987 article. Our impressions about the earth's past climate and how the climate system works are more than just educated guesses:

> They come from the results of experiments nature has conducted on her own. The results of the most recent of them are well portrayed in polar ice, in ocean sediment and in bog mucks. What these records indicate is that Earth's climate does not respond to forcing in a smooth and gradual way. Rather, it responds in sharp jumps which involve large-scale reorganization of Earth's system. If this reading of the natural record is correct, then we must consider the possibility that the main responses of the system to our provocation of the atmosphere will come in jumps whose timing and magnitude are unpredictable.

The big problem, Broecker went on to explain, is that the underlying mechanisms that amplify effects from changes in the composition of the atmosphere are still not well understood and therefore cannot be adequately modeled. The changes in the atmosphere occurring now, although comparable to those that happened during the ice ages and interglacial periods, are going in the opposite direction. That is, during the ice and interglacial ages, the carbon dioxide content of the atmosphere fluctuated between roughly 200 and 300 ppm. Today it is heading toward 400 ppm. Considering that when it was 200 ppm, civilized life in most of North America and Northern Europe would have been unsustainable, what will the impact on civilization be when it is 400 ppm or even higher? Could there be similarly catastrophic effects arising from parts of the world getting much too hot?

Because it is impossible to predict or even imagine what lies ahead, as quantities of greenhouse gases in the atmosphere approach levels never before recorded in the earth's geologic history, Broecker's description of the Younger Dryas event remains the most vivid and compelling example of the climate catastrophe that could happen later in this century if current trends continue unabated. Thus, in late 2002, when a report was issued by a scientific panel established by the U.S. National Research Council to evaluate the possibilities and dangers of abrupt climate change, the panel expressed worry about "an invasion of low-salinity deep waters that [have] spread over the entire subpolar North Atlantic Ocean and the seas between Greenland and Europe in just the regions critical for abrupt shifts in the thermohaline circulation." Though the panel reckoned the probability of catastrophic climate events in the near future to be low, it did not dismiss their possibility and drew sharp attention to the fact that in the earth's known climate history, changes way out of proportion to causes have occurred regularly: small, globally averaged forcings produced temperature changes during the 100,000-year ice age cycles of 5 to 6 degrees Celsius; and on shorter time scales, temperature changes about a third as big have occurred in 10-year periods. Such changes, were they to happen now, could have regional effects that would be difficult or impossible to adapt to. And, the panel warned, the modeling techniques on which current IPCC projections are based "have not...reached the level of sophistication that will enable them to be used to simulate the likelihood of occurrence of the more abrupt and possibly spontaneous climate shifts [that could produce such difficulties]."[17]

Taking a much more alarmist line in a report that got into the U.S. press in February and March 2004, an intelligence group in the U.S. Department of Defense outlined a scenario in which temperatures might suddenly rise 5 or 6 degrees in the Northern Hemisphere and drop by similar amounts in the Southern Hemisphere, causing persistent drought, winter storms, and high winds.[18] Self-avowedly the kind of worst-case analysis so beloved at the Pentagon, the report took its cues from Broecker's interpretation of the Younger Dryas and another shutdown of the North Atlantic conveyor thought to have occurred 4,500 years after that, about 8,200 years ago. The authors took satisfaction from the fact that in recent centuries, instead of slaughtering all their enemies the way people did in ancient times, now "states merely kill enough to get a victory and then put the survivors to work in their newly expanded economy." Yet "all of that progressive behavior [!] could collapse if carrying capacities everywhere were suddenly lowered drastically by abrupt climate change," they worried.

Curiously, though the Department of Defense authors play up the dangers of abrupt climate change far beyond what even alarmists like Broecker would endorse, they pay only lip service to the idea of reducing the danger of catastrophic global warming by cutting fossil fuel consumption. Instead they emphasize the desirability of being able to predict when catastrophe will actually occur and of preparing ourselves to adapt when it does—a cataclysm they seem to think is almost inevitable.

If we believe abrupt climate change to be a real possibility if current energy trends continue unabated, what reasonable measures might we take to minimize the odds? What Broecker proposes, in a nutshell, is that we adopt heroic measures to capture carbon dioxide from power plants and find ways to dispose of it so it does not get into the atmosphere. But for reasons that will be explored more fully in the next chapter, this idea of "sequestering" vast quantities of carbon dioxide does not represent an economically feasible method of sharply reducing greenhouse gas levels in the immediate future—though it could become an important strategy later in this century. This is where Hansen and his alternative scenario come in handy. Contrary to those who argue that progress to date has been inconsequential and that it would take thirty Kyoto treaties to make any real dent in the problem of climate change, Hansen points out that modest but significant gains already have been made, and that their impact has been much more than trivial.

Fifteen years ago, Hansen observes, chlorofluorocarbons were, after carbon dioxide, the most important greenhouse gases warming the world. Their near elimination under the Montreal protocol adopted in 1987 has significantly mitigated the buildup of greenhouse gases already, making the world not quite as much warmer today as it otherwise would have been. Next, Hansen suggests, we could attack the contribution black carbon is making to global warming by limiting the fine particulates emitted by diesel-powered vehicles—which also are a known serious health hazard. By going for such "no regrets" options, like burning less diesel fuel or burning less coal—policies that deliver better health, even if their impact on climate turns out to be disappointing or simply unnecessary—everybody comes out ahead.

To be sure, there are political dangers in Hansen's moderation, just as his critics feared when he first promulgated his alternate scenario. In fall 2002, his ideas about black carbon got valuable support in an article by Stanford's Mark Z. Jacobson, one of the leading experts on the subject. Going significantly further than Hansen himself, Jacobson reported computer simulations showing that more could be gained in the short run by limiting particulates from fossil-fuel consumption and associated organic matter than by putting any reasonable limits on carbon dioxide and methane emissions. But already a year before the appearance of that paper, President George W. Bush's White House got wind of it and requested a draft; Bush promptly gave a speech in which he took Kyoto to task for leaving out limits on particulates and instead insisting on greenhouse gas limits that would "have a negative economic impact, with layoffs of workers and price increases for consumers" in the United States.

Jacobson, just as promptly, posted material on his personal Web site taking issue with the president on both points. First, he said, it did not follow from the Kyoto Protocol's omission of soot that the agreement was meaningless: "the Protocol would be strengthened by including black carbon, but it is not flawed in the absence of its inclusion." Second, he reminded his electronic visitors, clean air regulations requiring drastic reductions in some pollutants have been implemented without any visible damaging effects on the U.S. economy.[19]

If there is a moral in that vignette, it's that if we are to make headway toward averting Broecker's unpleasant surprises by proceeding stepwise with the measures that promise the biggest near-term returns, in the spirit of Hansen, we need to avoid the kind of partisanship that has stymied effective action in the United States for the last ten years. The traditional pattern has been for Democrats and environmentalists to advocate conservation and renewable energy—but not say anything critical about the coal industry, a traditional Democratic stronghold—and for Republicans, linked to the oil and energy industries, to belittle the prospects for conservation and renewable energy sources, boost fossil fuels, and promote nuclear power. In fact, if growth in greenhouse gas emissions is to be kept at reasonably safe levels, elements of both parties' programs will be required.

Clearly, the United States, which uses energy on a per capita basis far more profligately than any other industrial country, can do much more to conserve. At the same time, countries like Denmark and Germany have shown in the last ten years that with proper government incentives, renewable energy technologies can be deployed at astonishingly high rates. But there are clear limits to how much can be achieved by such measures: more than half of U.S. electricity comes from coal, and in countries like China and India, electrical dependence on coal is even greater; everywhere, transportation systems run primarily on oil, and secondarily on coal-generated electricity. Coal and oil are not going to be replaced wholesale by renewable sources of energy in this century.

Sachs, the man charged by Columbia University with reconciling the conflicting views of scientists like Broecker and Hansen, prides himself on his skill at wielding what economists like to joke is their most powerful tool—long division. It is a tool the reader will need to use in assessing the potential of conservation and renewables and estimating the need for other sources of energy.

PART THREE

CHOICES

*The Low-Carbon and Zero-Carbon Technologies
We Can Deploy Right Now*

A wind turbine is readied for installation at Denmark's Horns Rev offshore wind farm, the world's largest.   *Source:* Robb Mandelbaum

CHAPTER 8

Breaking the Carbon Habit

S INCE THE MID- TO LATE 1990s, virtually all qualified scientists have firmly agreed that fossil fuel combustion is warming the world.[1] So strong is the consensus, journalists have begun to wonder whether their normal instinct to tell both sides of a story has actually led them to produce unbalanced work—giving skeptics about global warming much more leeway than they deserve.[2] The seeming unanimity among scientists is, in truth, somewhat deceptive. While there's general agreement that something ought to be done to rein in greenhouse gas emissions, if only as insurance against the possibility that global warming could have severe consequences, professional opinion varies widely about just how serious those consequences might be. What is the likelihood of changes in climate and ecosystems that are too rapid and too sharp for societies to cope adequately? How much action is called for now, in light of uncertainty about the severity of possible outcomes? What specifically should be done? On close inspection, it turns out that scientists don't even agree about the meaning of the word "likely," let alone the actual likelihood or improbability of future climate scenarios.[3]

For rather obvious reasons, having to do not only with scientific uncertainty but also with politics and rhetoric, defining "dangerous climate change" has been especially divisive. This key phrase is enshrined in the International Framework Convention on Climate Change, which was adopted at Rio de Janeiro in 1992 and has been ratified by pretty much all the nations of the world, including the United States. (The senior President Bush signed it.) The Framework Convention is the basis of the Kyoto Protocol, the implementation agreement drafted five years later, largely at U.S. behest, but then repudiated by successive

U.S. governments—first by the Senate, then, after the turn of the century, by the junior President Bush. The convention states as its general objective the prevention of dangerous climate change, and Kyoto is meant to start achieving that objective.

Many scientists avoid trying to define dangerous change, as any such attempt obviously involves value judgments. But policy makers naturally would like in principle to establish clear quantitative criteria: how much more greenhouse gas concentrations can be allowed to rise, how much of an increase in global mean temperature is tolerable, what kinds of changes in rainfall patterns are acceptable, and so on.[4] To the extent that experts express judgments on these questions, they range all over the map, with some, like Hansen, saying that any more than a degree of warming could be dangerous, while others, like Manabe, argue that the world ought to be able to adjust to several degrees or more, provided that the changes do not occur so fast that they make adjustment impossible and cause irreversible damage. But if Manabe is right that uncertainties in projections are destined to get larger rather than smaller in the near term, and that the near term could be quite long, the range of expert opinion is sure to remain wide and may get even wider. Under the circumstances, rather than attempt a quantitative definition of dangerous climate change, would it not be better to proceed on the basis of the fundamentals we know, taking what we don't know into account?

Even if global warming will be merely a matter of wet areas getting wetter and dry areas drier, of polar ice and mountain glaciers melting so that sea levels rise, and of storminess and weather variability increasing—the generally accepted Manabe–Wetherald scenario—it makes good sense to start limiting greenhouse gas emissions as a precaution. Those effects are already clearly visible and will become much more so. To slow and mitigate the known consequences of warming as much as possible, Kyoto seems a sensible first step: if the greenhouse gas emissions of the advanced industrial nations are reduced somewhat below 1990 levels, as Kyoto prescribes, then as emissions from rapidly industrializing countries like China and India increase, it might be possible to keep overall greenhouse gas concentrations from more than doubling in this century.

At a follow-up meeting to Kyoto held in Bonn, Germany, in July 2001, measures were taken to make the protocol more palatable to holdout countries, which at that point included Russia, Canada, the

United States, and Australia (Russia and Canada have since ratified the protocol). The Bonn agreements facilitated international trading in emissions credits, a system that reduces the overall cost of meeting Kyoto targets by allowing those who find it hard to meet the targets to purchase credits from others who are more easily able to meet and exceed them. Bonn also allowed countries to claim credit for the creation of carbon "sinks," the technical term for means of absorbing carbon, such as growing forests. But the carbon sink offset was limited in the crucial 2008–12 compliance period to a total of 55 million tons per year, barely 2 percent of total estimated emissions in that period, which are projected at 2.5 billion tons. That modest benefit, together with the boost given the emissions trading system, proved insufficient to lure the United States back into the Kyoto agreement and barely satisfied classically trained economists. William D. Nordhaus, the Yale University economist widely regarded as the foremost authority on the costs of carbon reduction, concluded that a "globally harmonized carbon tax"—a tax on carbon emissions, adjusted to reflect local realities—might have been preferable to Kyoto's carbon reduction targets as a method of cutting emissions. Nordhaus concluded, however, that it probably would not be possible to assemble a global coalition to rewrite carbon emissions tools; therefore, he found "major merit" in the Bonn–Kyoto approach: it's "the first experiment with market instruments in a truly global environment agreement." For that reason alone, Nordhaus said it is "hard to see why the United States should not join with other countries" in ratifying and implementing the accord.[5]

But would that be enough? Nordhaus estimated that even with the United States as a party, Kyoto would result in carbon emissions being only about 15 percent lower in 2075 than they would be if the world's business continued as usual, without carbon restraints. If the United States stays out, Nordhaus guesses that global emissions will be barely 1 or 2 percent lower than in the business-as-usual scenario—meaning, in effect, that the world will fast approach doubled carbon dioxide levels and be well on the way to a situation in which carbon dioxide is three or four times as high as in the preindustrial Holocene. Yet from the drillers, we know that there is a lockstep relationship between greenhouse gases and global temperature changes, and that in the geological past, relatively small changes in levels of carbon dioxide and methane—changes of the same order we are inducing today—were closely associated with the onset and termination of cataclysmic climatic events.

Even if increases in greenhouse gas levels continue to be very gradual and global temperatures rise in small, even steps, consequences still could be severe and, in some regions of the world, even catastrophic.

Already, serious effects from global warming are becoming apparent; scientists are notably less shy about attributing them directly to climate change than they were only a few years ago; and projections of future effects are becoming more credible all the time. Many of the effects can be found right at your front door, if, like Donald Kennedy, the former Stanford president who now serves as editor in chief of *Science* magazine, you happen to be an amateur naturalist. "Flowering seasons come weeks earlier for British plants than they did in the 19th century when parson-naturalists began recording the dates on which their favorite species first bloomed," Kennedy has observed. "Birds familiar to American birders have spread their ranges northward: Cardinals now winter in New Hampshire, a state they never reached in summer when I was a boy. In the intertidal zones of Monterey [California] the transects... are now occupied by species that had more southern distributions in the heyday of Cannery Row. And on the other coast bluefish and striped bass, unknown north of Cape Cod when I spent my first summer in Woods Hole, now delight anglers in the Gulf of Maine."[6]

Such evidence might be dismissed as the stuff of trivial sentimentality, but that certainly cannot be said of what is happening at the earth's poles, or of what's happening and could happen in some of its breadbaskets. At the North Pole, Arctic ice thinned by an average of close to 50 percent in the last decade of the last century, and scientists projecting what might happen next even have begun to entertain the possibility that the entire Greenland ice sheet might melt in this century, which would have a huge impact on global sea levels. According to the glaciologist and paleoclimatologist Richard B. Alley, the number of days oil companies are able to conduct operations in Alaska has fallen by half in recent years because of softening permafrost;[7] communities in Alaska and Siberia are seeing their foundations literally crumble. Almost as alarming are events at the South Pole, where "ice shelves the size of American states are either disintegrating or retreating," *The New York Times* has reported.[8] In 1995, the Larsen A ice shelf in Antarctica disintegrated, followed in 1998 by the collapse of the nearby Wilkins ice shelf. "Over a 35-day period early in 2002, at the end of the Southern Hemisphere summer, the Larsen B ice shelf shattered, losing more than a quarter of its total mass and setting thousands of icebergs adrift in the Weddell Sea."

In both Greenland and Antarctica, when ice melts abruptly where glaciers meet coasts, it is as if a cork were pulled from a bottle: glaciers begin to advance much more rapidly into the oceans. On the so-called Antarctic peninsula, which in fact is a collection of islands and glaciers, the discharge rate of three of the biggest glaciers increased eightfold just from 2000 to 2003. Closer to the pole, there is an even more vulnerable area where ice shelves are as big as Texas or Spain and the land they rest on is below sea level. As for the catchment region of Amundsen sea glaciers, if all its ice melted, it would raise the global sea level 1.3 meters.[9] According to Alley's estimates, if the Greenland ice sheet were to give way, oceans would rise 7 meters globally, and Miami Beach would have to move north of the Everglades; if West Antarctica were to go as well, oceans would rise 12 meters; and if both West and East Antarctica went, oceans would rise 70 meters and Florida would disappear entirely.

Plausible drastic consequences are by no means limited to polar or semitropical regions. In some of the world's temperate regions, too, climate change is already leading to disturbing effects that could, in time, become truly catastrophic. Persistent droughts have been sorely straining water resources, threatening agricultural output in the American Southwest, where summer temperatures now routinely reach 115 degrees Fahrenheit, and in north-central China, where the headwaters of the country's fabled Yellow River have been greatly affected. In some recent years, the river has gotten so low it doesn't quite make it to the sea. Those droughts may still be within the range of natural variability, but scientists are much more willing than they were ten or twenty years ago—when Hansen used the metaphor of loaded dice to explain the statistical relationship between weather events and climate change—to draw a direct causal link to global warming.

For example, several scientific teams have taken a close numerical look at the heat wave that ravaged Western Europe in the summer of 2003, causing an estimated 22,000 to 35,000 premature deaths in countries from France to Ireland.[10] A group at Switzerland's ETH constructed a bell curve mapping the probabilities of heat waves and found that the 2003 event was literally off the chart—much less probable than a 1,000-year heat wave and more like a million-year heat wave.[11] John F. Mitchell, chief scientist at Britain's Hadley Centre, used methodology borrowed from smoking studies to identify the purely human contribution to the risk of extreme events occurring. He and his collaborators found that the 2003 European heat wave could be unequivocally traced to the effects of fossil fuel combustion. (And when a critic dismissed the work as "small potatoes," Mitchell and his two Irish coauthors could not resist: they went back to the drawing board, gathered data on the size of Irish potatoes, and determined that they were indeed 30 percent smaller than average during the summer of 2003!)[12]

Looking ahead, scientists at Harvard University have estimated that as greenhouse gas levels rise, "the severity and duration of summertime regional pollution episodes in the midwestern and northeastern United States increase significantly relative to [the] present. Pollutant concentrations during these episodes increase by 5–10 percent and the mean episode duration increases from 2 to 3–4 days."[13] With warmer temperatures, the number of days when ozone levels are unhealthily high rises, photochemical reactions that produce smog speed up, biogenic emissions of allergens increase, and air stagnates, trapping pollutants and allergens.

Even more drastically, researchers at the University of Illinois, Urbana–Champaign, have estimated that if business continues as usual, by the end of the century Illinois will have a climate that closely resembles that of arid northeast Texas. That's certainly not small potatoes: Illinois and neighboring Iowa are one of the world's great breadbaskets, and the lucky farmers living in small towns in western Illinois have among America's highest average family incomes. Farmers in northeast Texas mainly grow—or try to grow—cotton on dry, unirrigated land.

All these kinds of consequences, which could follow linearly from small, steady, incremental increases in greenhouse gas concentrations, are serious enough. But from the computer modelers, we know that climate systems can oscillate between radically different states, and that small changes in variables can prompt a transition from one state to another. It is of course possible and indeed probable that as greenhouse gas levels double or triple or quadruple, no systemwide changes will occur; the effects may be confined to those that modelers like Manabe and Wetherald anticipate. But is there not also some possibility, as Broecker senses, that the great atmospheric changes brought about by current activities could cause some kind of catastrophic system change? And how should we think about such scenarios, which are real possibilities, even if they are highly improbable?

The fact that climate change could occur suddenly, perhaps with dire consequences, has gained currency in recent years as many general-interest science magazines have run feature articles on the subject of what's usually called "abrupt" climate change.[14] Perhaps this has been a bit too much of a good thing, in that it has led to self-parodying treatments like the one in *The Day After*, the 2004 movie in which Los Angeles is leveled by tornadoes, and New York City, in a matter of days, is buried under mountains of ice. Certainly the emphasis on abruptness has been slightly distracting, in that it's a question not just of how fast climate change might occur but also of whether the change would be drastic enough to threaten the foundations of civilizations.[15]

To judge from the behavior of the insurance industry, which will be significantly affected by climate effects, the message is only just beginning to sink in. Insurance is a three-trillion-dollar-a-year business globally, and its fortunes will depend on whether global warming produces damages that are merely incrementally worse than what one would project from the recent past, or whether they turn out to be drastically different in kind. Leading companies in the reinsurance business, like Munich Re and Swiss Re—companies that insure insurance companies against catastrophic losses—are taking the risks of cataclysmic climate change seriously indeed: they employ climate scientists as consultants and sponsor expert conferences to assess probabilities. But they are the exception: most insurance companies continue to rely mainly on historical extrapolations rather than dynamic climate models to assess climate risks, and if it turns out they are greatly underestimating future damages because they rely too much on past patterns—as will likely be the case—the general public will end up paying the price. Even under normal circumstances, private insurance covers only a quarter or a third of the damages resulting from disastrous weather.[16]

One person who *has* gotten the message, evidently, is the genius financier Warren Buffett, who is reputed to have said, "Catastrophe insurers can't simply extrapolate past experience [anymore]."[17] Another is Richard Posner, the brilliant Chicago-based federal judge who writes on a wide range of subjects of public interest, from law and public policy to the impeachment of former president Bill Clinton (which he considered justified). Much influenced by the laissez-faire University of Chicago school of economics associated mainly with Milton Friedman, Posner is not naturally prone to consider a problem ripe for government intervention, or even necessarily to classify problems as matters of public interest at all. But in a 2005 book, *Catastrophe: Risk and Response*, Posner counts climate change among a number of major disasters that he believes could drastically affect the world and that are not yet taken as seriously as they should be.[18]

Posner's general view is that when the risk of a catastrophe is small or remote, we have trouble acknowledging its reality at all. In a way, this is the flip side of a well-known and often-discussed phenomenon sometimes called the lottery problem—in hope of winning huge sums of money, people often prefer to make wagers against ridiculously long odds, rather than go for much smaller sums with enormously greater chances of success. In essence, we all find it hard to make rational choices where very small probabilities are involved. Nonetheless, as Posner wrote in a *Wall Street Journal* column following the 2004 tsunami that ravaged much of Southeast Asia, "The fact that a catastrophe is very unlikely to occur is not a rational justification for ignoring the risk of its occurrence. The risk may be slight, but if the consequences, should it materialize, are great enough, the expected cost of disaster may be sufficient to warrant defensive measures."[19]

Considering this problem of how to make rational decisions about remote possibilities, there is an instructive moral in the chain of events that led to the construction of the first atomic bomb. It involved a noted exchange between Leo Szilard, the Hungarian physicist who was the first to see clearly how a bomb could be made, and Enrico Fermi, the Italian physicist who presided over the design and construction of the first reactor in which a self-sustaining nuclear chain reaction took place. The communication took place at Columbia University, shortly after the discovery in Berlin of nuclear fission and on the immediate eve of World War II, with the young Columbia physicist Isidor Rabi acting as intermediary.

Szilard and Fermi were in some respects polar opposites: Szilard was a social visionary who wandered about from country to country and always, by his own account, considered it his personal mission to save the world; Fermi, in contrast, was a rather apolitical person who lived and breathed physics, morning, noon, and night. Their conversation, reported in an absorbing memoir by Szilard, went roughly like this: Rabi told Fermi that Szilard considered the construction of an atomic bomb possible, with the implication that if Nazi Germany got one first, it would be big trouble. Fermi said the idea was "nuts." But when Rabi reported Fermi's reaction to Szilard, Szilard asked why Fermi thought the idea was so silly, and the two went back to Fermi to find out. Fermi replied that there was only a "remote possibility" of an atomic bomb being made successfully. When Rabi asked what he meant by "remote," Fermi said perhaps a 10 percent chance. To which Rabi then said, "Ten percent is not a remote possibility if it means that we may die of it. If I go to the doctor and the doctor tells me there's a remote possibility that I might die, and that it's 10 percent, I get excited."[20] Whereupon Fermi got excited, and proceeded to lay the foundation for the U.S. atomic bomb with the construction of the Chicago reactor.

Rising carbon dioxide emissions call for similar resolve. Though the likelihood of their causing some kind of catastrophic climate change may be considerably smaller than 10 percent, perhaps even less than 1 percent, the dangers associated with that change are potentially huge. Concerted measures need to be taken right now, not just to slow the rate of gas buildup but to reverse it. The longer the United States, in particular, delays, the more drastic the measures will be that are necessary to strike a new course. And it's no use complaining that Kyoto puts unfair burdens on the United States, as the Bush administration has claimed. To begin with, considering that per capita U.S. greenhouse gas emissions are about 5 times Europe's and perhaps 20 to 25 times China's or India's, the fairness argument doesn't have much merit on the face of it. But that's not really the point. When confronted with a possible disaster that can only be averted by resolute action, you don't procrastinate on grounds that somebody else is not doing their fair share. You do what you can, with all dispatch, and hope that as many others as possible will soon join the effort.

Again, World War II provides an instructive moral. Before U.S. entry into the war, when Franklin D. Roosevelt sought to persuade the American public of the wisdom of his Lend-Lease program—to "lend" equipment and materiel to Great Britain, without being too fussy about when it would be repaid—he conjured up the image of a neighbor's house on fire, endangering yours as well. If your neighbor came over to borrow a garden hose, you wouldn't quibble about the $15 it cost you. You'd give your neighbor the hose, hoping that you'd eventually get it back—and that maybe, someday, you'd get a return favor as well.[21]

As greenhouse gases are reaching levels far, far higher than anything experienced in the last 700,000 years, the possibility of something catastrophic happening to earth's ocean-atmospheric system, however remote it may be, cannot be ruled out. Perhaps the probability is even smaller than the chances were of Hitler's building an atomic bomb successfully (he was nowhere near). In any case, the probability is not zero and not trivial. Meanwhile, however, it's as if we and our neighbors were pouring combustible fuels onto all the homes on our block, disregarding the growing possibility that they might catch fire. Instead of just ceasing and desisting as fast as we can, even though we are adding by far the most fuel, we have been arguing about whether some of our neighbors are stopping fast enough.

So what can the United States do? The situation calls for using all available technology already proven to reduce carbon emissions right now, and that means first and foremost finding ways to radically cut both gasoline consumption and coal combustion. Each accounts for about one third of U.S. carbon emissions, so without making deep cuts in each, no serious program of carbon abatement is conceivable. It's generally acknowledged that burning gasoline in cars results in carbon emissions, and that the only way of preventing those emissions is to drive less, drive more fuel-efficient cars, or power vehicles some other way altogether. The question of just how much benefit can be hoped for from each of those alternatives—particularly the much-trumpeted "hydrogen economy"—will be addressed in chapter 10. As for coal, the answer is to phase out its use, conserving electricity as much as possible and substituting zero-carbon and low-carbon fuels, including both renewables and nuclear power (the subjects of chapters 9 and 11).

The coal and utility industry would have the American public believe that coal is part of the solution to global warming, as well as to U.S. dependence on foreign oil suppliers, not part of the problem. But in a near-term perspective—the point of view taken in the remainder of this book—the industry's claims do not withstand critical scrutiny. Synthetic fuels extracted from coal are not competitive with gasoline, even at the relatively high oil prices of 2003–05, so coal is not a substitute for imported petroleum. In terms of climate, coal has a useful role to play only if the carbon liberated in its combustion can somehow be captured and stored—"sequestered," in the jargon of the trade. What causes confusion is the notion of "clean coal," often the subject of "public service" messages on radio and television. In fact, as explained in chapter 2, coal scrubbed of its sulfates is worse than dirty coal with regard to global warming, because the sulfates reflect radiation and counteract the warming effects of carbon dioxide. The only way to make coal clean in a climatic sense is to sequester the carbon dioxide. But sequestration, though it has considerable long-term promise, cannot do much in the decades just ahead to sharply reduce carbon emissions from coal-fired plants.

Although carbon dioxide can be and is extracted from power plant emissions for specialized applications, and although it can be stored by various means under the sea or in geological repositories or even injected into oil and gas wells to enhance recovery, these complex processes are very expensive and risky when attempted on a large scale. To make a serious dent in carbon emissions from coal burning, hundreds of millions of tons would have to be stowed away annually. The quantities involved are similar to the quantity of petroleum moved around the world in the present oil economy,[22] which shows that the creation of a carbon-capture economy is certainly possible, but also a very expensive proposition.

Howard Herzog, a professor at the Massachusetts Institute of Technology (MIT) who has been one of sequestration's most ardent proponents, has estimated, perhaps a little too optimistically, that capturing and storing carbon dioxide emissions from power plants would boost electricity generation costs by two or three cents per kilowatt hour.[23] That might not sound like much, but in fact it's close to the cost of producing electricity from the most efficient natural gas-fueled turbines or of making it from coal. So sequestering all the carbon emissions from fossil fuel-powered plants might double the average "busbar" cost of electricity—the cost of making it alone, not counting the (higher) costs of transporting, delivering, and billing for it.

Problems with sequestration go beyond the narrow issue of money. If done on a large scale, there also are serious concerns about long-term stability. If enough sequestered carbon dioxide were to suddenly well up from subsurface reservoirs in a populated area, it could asphyxiate anybody in the immediate vicinity. The prime example of what could happen is a natural tragedy that occurred in 1986, in the African country of Cameroon. A plume of naturally present carbon dioxide abruptly bubbled to the surface of a volcanic lake and choked an estimated 1,700 people to death. Taking such risks and the potential liabilities associated with them into account, an MIT–Cambridge University study concluded that although they have "received little attention in the literature, [they] could significantly affect the viability of carbon storage as a long-term solution to climate change."[24]

From a purely technical point of view, and with the motivation of tax incentives, there's no doubt that carbon sequestration can contribute to greenhouse gas abatement. Herzog observes that if sequestered carbon benefited from the same 1.7 cents per kilowatt-hour tax credit that wind energy has had, "you will see quite a bit of carbon sequestration."[25] Since 1996, spurred by a stiff Norwegian tax on carbon dioxide emissions from offshore oil drilling, Norway's Statoil has compressed and pumped large quantities of carbon dioxide into a 200-meter-thick subsea sandstone formation near an oil and gas field 240 kilometers from shore. Statoil has stashed away about a million metric tons of carbon dioxide annually in the Sleipner Formation, at an initial investment cost of $80 million—all of that recovered in less than two years from avoided carbon-emission taxes.

In North Africa, in a remote part of Algeria, a consortium led by British Petroleum has invested $100 million in special equipment to pump carbon dioxide released in natural gas drilling operations back into the gas reservoirs. The objective, so far thwarted by technical setbacks, is to sequester about a million tons of carbon dioxide each year, roughly half the amount normally released in the extraction process.[26] In terms of carbon savings, that's equivalent to doubling the fuel efficiency of 500,000 cars, building 500 one-megawatt wind turbines, or covering 1,000 acres with photovoltaic panels—but it's still a drop in the bucket compared to total world emissions, it's not cheap, and it's not hassle-free.

In North America, carbon dioxide has been run from a coal gasification plant in North Dakota through a 205-mile pipeline to a town in Saskatchewan, where an oil company pumps it into the ground to enhance petroleum recovery. The Dakota gasification complex in Beulah originated in the "synfuels" program President Jimmy Carter launched in the 1970s, as part of his energy independence initiative. Synfuels, based largely on technology Germany developed during World War II, was a near total flop, commercially. But at least the Canadian end of the Dakota project helped demonstrate the feasibility of sequestering carbon dioxide for oil recovery.

No doubt sequestration, together with clean coal and coal efficiency improvements, will have an important role to play in the long run, as technologies mature, especially in countries like China and India that have no choice but to keep burning coal in huge quantities for the foreseeable future. But it is simply wrong to suggest, as some continue to do,[27] that carbon capture and storage represent the most cost-effective means of cutting carbon emissions in countries like the United States, or that sequestration technologies are ready to be deployed on a very large scale right now.

The U.S. National Research Council and the National Academy of Engineering have sponsored a number of workshops and conferences to explore innovative methods of carbon capture and storage. The details of presentations clearly show that both economic and technological hurdles are enormous. For example, a February 2003 workshop was convened "to identify promising lines of research that could lead to a currently unforeseen breakthrough in the development of carbon-free energy systems."[28] Besides injection of carbon into oil and gas reservoirs to enhance recovery, the workshop explored ideas like injection into deep ocean sediments, where pressures and temperatures would keep carbon dioxide at a density greater than water, so that it would be isolated from seawater and therefore not subject to undesired chemical reactions. Alternatively, carbon dioxide might be injected into subsea areas where tectonic plates are separating, or stored chemically in hydrates found below the permafrost region in Arctic areas. An approach favored by Klaus Lackner at Columbia University, "chemically enhanced weathering"—specifically, the acceleration of carbonation in magnesium silicate—was found to be too energy-intensive to be practical.[29] (The workshop suggested that perhaps bacteria or microbes could be found that would better promote weathering of serpentine to absorb carbon.)

Summing up, the workshop recommended that the Department of Energy require any project proposals for carbon capture and storage to show that "the scale of carbon sequestration achieved is commensurate with the problem, that the thermodynamics are real, and there is some consideration of cost."[30] In other words, the project organizers should show some appreciation of the enormity of the problem and suggest things that realistically might work; potential economic viability of an approach would be a consideration but not a requirement.

A symposium sponsored by the National Academy of Engineering the year before, in April 2002, had a similar tone. Robert H. Socolow of Princeton University asked whether we should strive to develop storage systems that, like proposed nuclear waste storage facilities, "future generations can undo." He wondered, "How will we keep the overall costs of storage from escalating to the point where the prognosis for the whole strategy becomes bleak?"[31] Franklin M. Orr of Stanford University took note of the inconvenient disjunction between where carbon dioxide generally is produced and where it might most effectively be stored. "For CO_2 sequestration to be practical ... we will have to put forth significant effort. We must also expect that many approaches will be part of the solution. There will be no silver bullet."[32] Gardiner Hill of the British Petroleum Group, noting that three quarters of sequestration costs typically are associated with capture and one quarter with sequestration, observed that "overall, the cost of capture and storage today are very high," so "we are always looking for new ideas."[33] Reporting on a pilot enhanced oil-recovery project mounted by Chevron and Texaco in California, Sally Benson of Lawrence Berkeley National Laboratory said that the project became "very challenging, particularly in light of what we were trying to do."[34]

In short, for carbon sequestration to make a really big dent in atmospheric carbon dioxide concentrations, a very great many problems will have to be solved in tandem. Economically viable means of monitoring storage sites for leaks have to be found, as well as more efficient ways of conducting global surveys for suitable sites. Using current methods over hundreds of years, said BP's Hill, would "add prohibitive costs."[35] It may be true, as Herzog says, that with a carbon tax of, say, 2 cents per kilowatt-hour, quite a lot of carbon sequestration would be done. But other effects would be much greater, at least in the first decades such a tax were in place.[36] The coal and utility industries have been arguing for years that if they are subjected to very strict pollution controls and then have to pay carbon taxes on top of pollution abatement costs, they might as well just shut down many plants immediately. The total costs of meeting the air quality standards adopted by the Bush administration in early 2005, the so-called interstate rule, are estimated at $50 billion.[37] If a carbon tax were set high enough to have some real bite—say at 3 cents per kilowatt-hour—it would indeed lead to much of the U.S. coal industry shutting down. This is why climate-conscious investor groups increasingly have demanded that coal-burning utilities present concrete plans to shareholders explaining how they are going to move away from coal. The utilities also are under continuing and mounting pressure from state and municipal authorities to address not just pollution concerns but carbon abatement as well.

If a carbon tax were set high enough to result in much of the U.S. coal industry being shuttered, and if that coal-generated electricity were replaced by zero-carbon fuel-generated electricity, the net effect would be to cut U.S. greenhouse gas emissions by one quarter to one third—more than required by Kyoto, and not unaffordable. But how realistic is it to suppose, in the "no new taxes" political environment of the United States, that some kind of carbon tax might actually be imposed? At this writing, it's obviously utterly unrealistic. But political realities can change, and they must be changed. As the late diplomat and diplomatic historian George F. Kennan famously said, a country that shortchanges its global responsibilities because of supposed domestic political realities is a country that invites catastrophe: "History does not forgive us our national mistakes because they are explicable in terms of our domestic politics."[38]

Let us assume, for the sake of argument, that Americans can be persuaded by a more visionary political leadership that the threat of climate catastrophe is real and that a concerted effort must be made immediately to reduce greenhouse gas emissions by roughly a quarter. What technologies are available right now to be used in such an effort? Carbon sequestration, as discussed, may be promising in a longer perspective but is not ready to play much of a role in the next decade. What about renewable energy technologies like wind and solar—how much can be expected from them in the United States? And what about energy conservation and improved energy efficiency?

There has been much talk in recent years about the so-called "hydrogen economy"—the notion that vehicles powered by batterylike fuel cells could soon replace cars and trucks running on internal combustion engines. Will that happen in the next ten or twenty years? If not, should there be more emphasis right now on tightening fuel efficiency standards or perhaps raising gasoline taxes? Let's say a visionary leadership imposed a flexible gasoline tax system, so that prices were kept high enough to guarantee investments in developing alternatives to the internal combustion engine. How high should such a tax be, and how much could it be expected to reduce gasoline consumption and greenhouse gas emissions in the near term?

Finally, even with an all-out effort at deploying renewable technologies, conserving energy in industry, commerce, and residences, and cutting total fuel consumption, will greater reliance on nuclear energy also be necessary? Can the United States accomplish a near-instantaneous 25 percent cut in greenhouse gas emissions without building any new nuclear reactors?

CHAPTER 9

Going All Out for Renewables, Conservation, and Green Design

WHEN WORLD ENERGY DEMAND is projected for the next half century and estimates are made of how much low- or zero-carbon energy would have to be substituted to prevent greenhouse gas emissions from more than doubling, the figures are at first glance so gargantuan, the situation seems almost hopeless. The International Energy Agency in Paris has concluded that if business continues as usual, with no special efforts made to curtail energy demand, it will grow in just the next 30 years by two thirds; fossil fuels will supply 90 percent of that increase.[1] The IEA projects that electricity demand will double in the same period, because as countries become more advanced economically, they tend to substitute clean and convenient electricity for more primitive energy sources. Meeting the growth in electricity demand would require adding 1,500 gigawatts (billion watts) in new generating capacity. That's roughly equivalent to the construction of 1,500 standard nuclear power plant complexes, a daunting and unsettling prospect: if that added electricity demand were met by nuclear reactors alone, one nuclear power plant complex would have to be built somewhere in the world every week for the next three decades.

The Energy Agency's business-as-usual forecast implies of course that greenhouse gas levels will be much more than twice preindustrial levels by the end of this century. It's easy to see why the crude numbers suggest that the task of keeping levels significantly lower than that seems hopeless, especially inasmuch as greenhouse gas levels thirty, fifty, or one hundred years in the future depend not just on emissions at those times but also on virtually all emissions in the meantime. Accordingly, bringing future levels down requires doing much more than just having low-carbon energy sources supply increases in demand. Stabilization of gas levels requires concerted action right now to sharply reduce emissions, and this is generally recognized by both greenhouse skeptics and greenhouse alarmists. This essentially is why the Bush administration has taken the position that the United States cannot afford to do much to reduce emissions, and that even if the United States did try to meet Kyoto requirements, it wouldn't make much of a dent in the alleged problem. And this is also why some scientists, like NCAR's Jerry Mahlman, have said it would take many, many Kyotos to make a significant difference.

Teams of U.S. researchers have concluded independently that holding greenhouse gas levels at no more than 550 ppm, roughly double their preindustrial levels, would require deployment of 15 terawatts (trillion watts) of emission-free energy sources by mid-century. Taking into account that 3 thermal watts translate on average to one electrical watt, the amount of carbon-free energy required is more than three times total electricity growth in the IEA scenario. Members of one team, led by Martin Hoffert of New York University and Ken Caldeira of Lawrence Livermore National Laboratory, have characterized the challenge as "Herculean." They see it as requiring nothing less than a NASA-type research program to accelerate commercialization of technologies like nuclear fusion, satellites designed to beam solar energy back to earth, carbon sequestration, and superconducting grids.[2] The trouble with this kind of all-out research and development program, however, is that it really does amount to throwing money at the problem: though technologies may seem very promising, there's no way of telling in advance which ones will pan out, or when; and accordingly, there's no way of being certain that the envisioned breakthroughs will actually contribute to solving problems within the required time frame. For example, a commercial breakthrough in fusion has been predicted to be about 25 years away for at least 25 years now—and many observers have come to feel that, like a mirage, it may always be 25 years away. The solar satellite concept, though technically intriguing, has come in and out of fashion repeatedly in just as long a time.[3] Yet Hoffert and Caldeira conclude that without some such breakthroughs, stabilization of greenhouse gas levels will be simply impossible. They explicitly reject the IPCC's contention that "known technological options could achieve a broad range of atmospheric CO_2 emission-free power." Energy sources capable of stabilizing emissions "do not exist operationally or as pilot plants."

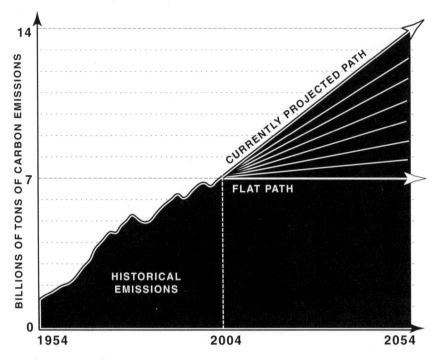

Carbon Mitigation "Wedges"
Source: Carbon Mitigation Initiative

Researchers at Princeton University's ecology department and engineering school, Stephen W. Pacala and Robert Socolow, emphatically disagree with that conclusion.[4] Anticipating, like Hoffert and Caldeira, that in a business-as-usual scenario carbon emissions will rise from about 7 billion tons a year now to about 14 billion tons a year 50 years from now, Pacala and Socolow ask how that doubling can be prevented. To break the monster problem down into more digestible pieces, they divide the 7 billion tons of carbon emissions to be avoided by mid-century into what they call "stabilization wedges." Each of the wedges represents a demonstrated activity that could reduce emissions, increasing its effect linearly over 50 years, so that by mid-century it would account for a billion tons of reduced emissions (see graph). Thus, 7 such wedges would reduce emissions by the required amount.

Pacala and Socolow identify 14 technology wedges that each have the potential to reduce carbon emissions by a million tons a year by mid-century, and basically invite their critical readers to pick 7. For example, emissions from oil combustion could be reduced one million tons by requiring automobiles to be more efficient or by inducing people to drive fewer miles. Specifically, the average fuel economy of 2 billion cars could be steadily increased from 30 miles per gallon to 60 miles per gallon, or each of those cars at current gasoline consumption rates could be driven 5,000 miles per year rather than 10,000. Alternatively, natural gas or nuclear energy could be substituted for coal at electric plants: replacing 1,400 one-gigawatt coal plants with natural gas plants would yield the desired million-ton reduction in emissions; replacing 700 with nuclear reactors would yield the same carbon savings. Capturing carbon dioxide at hydrogen production plants, coal-fired power plants, or plants producing synthetic fuels could each yield a million-ton emissions wedge, but it would take 3,500 Sleipner-scale geologic storage facilities to contain the carbon dioxide retrieved by any of those methods. Boosting the average efficiency of coal-fired power plants from 32 percent to 60 percent by replacing old technology with new could produce a million-ton emissions saving; and so, too, could improving the efficiency of buildings and appliances by just a quarter over a 50-year period. If wind turbines were substituted for coal-fired generation, the current number of turbines would have to be multiplied by 50 to produce a one-gigawatt wedge. Substitution of photovoltaics for coal would require current solar capacity to be increased 700 times.

Socolow and Pacala are well aware that some of the emissions savings from their candidate wedges will happen anyway in business-as-usual scenarios, and they appreciate that not every wedge will seem plausible or desirable to every critical reader. But they do show convincingly that even if the reader accepts only half of what they propose, the problem of greenhouse gas stabilization can in principle be solved. It's really just a question of deciding which fixes are most acceptable or least offensive, and then systematically adopting them as soon as possible. In other words, kicking carbon means acknowledging that of the methods available to get the job done, certain specific ones seem on balance to be most acceptable.

When it comes to deciding which low- and zero-carbon technologies should be used to reduce future carbon emissions, surely just about everybody's first choice will be renewable sources of energy like wind and solar power, and improved efficiency. The advantages of wind and solar are obvious—no depletion of scarce and irreplaceable fossil resources, virtually no pollutants, and no carbon emissions. As for their disadvantages, whatever they are, they scarcely bear comparison with the death, disease, and environmental blight associated with coal, or the risks associated with catastrophic reactor accidents, nuclear waste disposal, and nuclear proliferation.

In the whole field of renewable energy, which in principle also includes hydroelectricity (although big dams are anathema to many environmentalists) and hydrogen separated from water (which some environmentalists have eagerly embraced), the really sensational story of the last decade has been wind. Encouraged by government laws that provide a modest production subsidy for every watt generated, there has been a tremendous wave of wind turbine construction across Germany's northern plains, in Denmark, and in Spain. England is following suit with an ambitious plan to girdle the isles with offshore wind farms, and only somewhat less ambitious programs are found from Texas to Spain and India.

The countries of the European Union added nearly 6 gigawatts of wind capacity in 2004 alone, bringing their total installed capacity to 34.1 gigawatts—nearly three quarters of global capacity.[5] And Europe appears to be far from the limits of what can be done to exploit wind. Although renewable energy has a general reputation for being merely a niche player, wind already is much more than that. Germany's turbines have a combined capacity approaching 18 gigawatts, the equivalent of more than 18 very large central power plants. This rapid buildup of wind power has depended on subsidization of turbine installations with small fees that are distributed among all German electricity consumers according to a formula set in the country's energy law. An expert energy advisory organization to Germany's Socialist-Green government concluded in early 2005, 6 months before its electoral defeat, that the country can still double wind capacity, which would enable the government to keep its commitment to phase out nuclear power without resorting to more coal.[6]

In Denmark, wind turbines provide at present about one sixth of the country's electricity, and the government intends to boost that share to 50 percent by 2030. The country's leading turbine manufacturers supply close to half of world demand for windmills, and they have an economic heft in Denmark itself comparable to the automobile industries in much larger countries. The nation's beautiful countryside is dotted everywhere with very tall turbine towers, mounted with whirring three-blade rotors, which don't seem to bother anybody very much. Nor has there been much controversy over very large wind farms—the world's largest, in fact—in Copenhagen's harbor waters and just off the southwestern coast of Jutland, the peninsula that juts up from the north German plains. Both complexes had somewhat costly problems, which have been damaging to the reputation of the leading Danish manufacturers, but the people's faith in wind seems unshaken.[7]

At the end of 2003, stealing a march on the Danes and Germans, Britain's Crown Estate—the organization that manages a large chunk of the monarchy's far-flung holdings, including its offshore continental shelf—announced the winners of contracts to build 15 wind farms at 3 locations off Great Britain's coasts. The wind farms, with hundreds of turbines each, will have a total generating capacity of 5.4 to 7.2 gigawatts and cost, in all, upwards of 7 billion British pounds (about US $13 billion). The project, said the UK's energy minister, Stephen Timms, put Great Britain on course to be producing 10 percent of its electricity from renewables by 2010 and 20 percent by 2020, compared with 3 percent now. It represents the second round of windmill construction spearheaded by British energy authorities.[8]

Britain's new turbine towers, which will be about 80 meters tall, are to be installed in 3 main areas: the river Thames estuary; the Greater Wash, 30 to 40 kilometers off the Lincolnshire coast; and the North West, extending from the north Wales coast to the Solway Firth and out into the Irish Sea. Developers were granted leases of 40 to 50 years. They include Warwick Energy Ltd., Airtricity (an Irish renewables company), the construction group Amec, Powergen (owned by Germany's EON AG), RWE Innogy (owned by the German utility RWE), and the oil companies Total and Royal Dutch/Shell. GE Wind Energy, a growing presence in Europe and worldwide, already is building a big farm at Gunfleet Sands off the Essex coast, as part of the first round of windmill construction.

It was an important vote of confidence for wind when General Electric, far and away the world's leading manufacturer of electrical equipment, got into the business in a big way when it acquired a windmill division from the shards of Enron, at what was probably a bargain-basement price.[9] The move signaled that wind power is on the cusp of full commercial viability, an assessment confirmed by calculations by MIT professors Richard K. Lester and John M. Deutch. Lester and Deutch, using a standard cost-benefit analysis that takes into account the interest that will be paid on any capital investment over the operating lifetime of the investment project, found that wind installation would be fully repaid after 10 years at a discount rate of 7 percent per year.[10] While modest subsidies still are required to encourage robust wind construction, this is because the technology is still relatively new and fast evolving, and risks are therefore difficult to assess perfectly.

In principle, anyplace where there's a lot of coastline with good prevailing winds, or where there are plains or gently rolling hills in reasonable proximity to centers of electricity demand, there is great potential for wind. And so, in a global perspective, there's no reason why wind could not be scaled up by a factor of 50 over 50 years, as Socolow and Pacala suggest. Conceivably the growth could be quite a bit faster than that. But there also are limits. Those most commonly mentioned are wind's intermittency, uneven distribution in space, and low power density—that is to say, the very large surface area consumed by turbines, compared to the amount of land covered by a central power plant producing electricity at an equivalent rate. Power densities depend enormously on the characteristics of the site and can vary from 2 to 15 watts per square meter, according to Vaclav Smil, a specialist on energy and the environment at the University of Manitoba. Often the best wind isn't where it's needed: in the United States, for example, North Dakota has enormous wind resources, but to get any electricity generated to major demand centers would require construction of very long high-voltage, direct-current lines, an expensive and environmentally dubious proposition. As regards wind's notorious fickleness, Smil—who basically is a strong advocate of renewables—has pointed out that in one major Texas wind farm, in the first 3 months of 1999, electricity generation was 20 or 30 percent higher than planned in January and February but then, in March, 70 percent below targeted production.[11]

Electricity only is produced when the wind blows, while demand follows daily and seasonal cycles entirely their own. Therefore, either some way must be found of storing electricity when winds are strong but demand is weak—which may not be possible on a large scale with present-day technology—or wind power must be backed up by conventional generators. The problem is mitigated in Denmark by proximity to major industrial centers in Sweden and Germany, so that Denmark can sell excess electricity into their grids or purchase electricity from them as needed. But not every country or region with good winds is so fortunate. If sources of wind are widely separated from centers of demand, not only does it cost more to get electricity back and forth, but grid-control problems also become significant. Over long lines, voltage tends to sag and alternating current cycles get out of sync, so that extra sources of electricity have to be built into the system if the lights are to be kept on reliably. While considerable progress has been made in recent years in addressing those problems,[12] even in a country as compact and happily situated as Denmark, the difficulties are not minor and are considered by local utility experts to be not fully solved.

To take account of issues connected with intermittency and grid control, Socolow and Pacala estimate that when wind is substituted for central power plant generation, 3 times as much wind capacity must be deployed as the capacity of the generation replaced. In effect, to make one of their billion-watt wedges, 2 million one-megawatt turbines would have to be built over the next 50 years, and they would consume roughly 30 million hectares—equivalent to 3 percent of the United States' land area. Smil guesses, not too inconsistently, that generating 20 percent of U.S. electricity from wind would take 1 percent of the nation's land—but only 5 percent of that land would actually be occupied by turbines. Though environmentalists and preservationists often object to wind turbines on aesthetic grounds, it's a positive attribute of wind energy that turbines do not much interfere with surrounding activities, such as commercial fishing, recreation, farming, and grazing.

The same can't be said for photovoltaics, which, at present levels of technology, have very poor power densities and consume huge physical areas for the amount of energy they produce, without those areas being usable for much of anything else. Since energy produced from photovoltaic (PV) cells is also, like the wind, highly intermittent—there's none at night, of course, and little on cloudy or foggy days—PV power also must be backed by generators that can be turned on and off as desired. Thus, as with wind, Pacala and Socolow require, if solar energy is substituted for centrally generated electricity, 3 solar watts of capacity for every watt displaced. To make one of their one-gigawatt wedges using current technology, 2 million hectares would have to be paved with solar cells—a factor of 15 less than for equivalent wind power—but those PV-paneled hectares won't be doubling as pastures or sailboat slaloms.

Solar power, so far, is one of those irritating glass-half-full, glass-half-empty stories—or maybe two-thirds empty, one-third full. Look at the charts one way, and progress seems impressive. In the 1980s and 1990s, prices for photovoltaic modules dropped from more than $23 per watt to less than $5.[13] The delivery price for PV-generated electricity came down, in just one decade, from 40 to 75 cents per kilowatt-hour in 1991 to 12 to 20 cents per kilowatt-hour, and in another decade, by some estimates, it may fall as low as 4 cents.[14] Efficiencies for the standard single-crystal and polycrystalline silicon cells exceeded 20 percent by the end of the 1990s, creeping steadily upwards toward their theoretical maxima, though module efficiencies were only about half that high.[15] Meanwhile, efficiencies for some of the materials envisioned as the stuff of next-generation cells crossed the 10 percent threshold considered essential for commercial viability: the ratio of generated electricity to the power of incident light in copper-indium-diselinide and cadmium telluride went from the vicinity of 5 percent in 1975 to 16 to 18 percent in 2000; in thin-film amorphous silicon cells, it went from less than 2 percent to nearly 14 percent in the same interval.[16]

Looked at from another perspective, however, the progress in PVs is a far cry from what many in the field expected when solar R & D began in earnest after the oil crisis of 1973. As the new century dawned a generation later, electricity produced by solar cells was still far from being price-competitive with electricity from the usual central generating sources; participants in international renewable conferences began to note, anxiously, that even after several decades of concerted research, thermal systems of converting solar energy to electricity continued to be more efficient and cheaper than systems based on photovoltaics.[17] The thermal systems, which can consist of parabolic troughs or dishes, typically concentrate incident sunlight on a fluid, which converts to steam and drives a standard electricity turbine. Yet even those systems have yet to demonstrate their commercial viability—the company that built the biggest pilot installation, in the Mojave Desert near Barstow, California, went out of business.[18]

Even more worrisome was the failure of the second-generation materials to actually be developed on schedule. Among specialists following the field, it was always taken for granted that the potential for first-generation silicon PV cells would be limited, because those cells have to be manufactured to a very high degree of purity using clean-room techniques borrowed from the semiconductor industry. Unless materi-

GLOBAL WIND AND SOLAR MARKETS, 2004

	Combined Power (MW)	Total Sales Value (millions of U.S. dollars)
Wind turbines	8,050	$7,245
Photovoltaic cells	1,050	$7,350

Source: Steven Taub, Cambridge Energy Research Associates

Per unit energy, photovoltaic cells in 2004 were eight times as expensive as wind turbines. PV power is not commercially viable in most situations, and thus not relevant to the issue of how to immediately reduce greenhouse gas emissions.

als could be developed that could be rolled out, extruded, or stamped into cells, it was hard to imagine thousands, hundreds of thousands, or millions of hectares being covered with them. Despite improvements in efficiencies, most of the materials considered as candidates for the second generation had drawbacks that stubbornly refused to go away—toxicity, a reluctance to bond properly with other needed materials, and problems with workability. Particularly disconcerting was the November 2002 decision by BP, the British oil company that billed itself for a while as the world's leading solar manufacturer and promised in its ads to go "beyond petroleum," to terminate U.S. production of amorphous silicon and cadmium telluride cells.[19] Hopes for next-generation PV turned to exciting developments with a new class of organic materials, with one U.S. startup in particular attracting some top scientific talent.[20] But the fact remained that the organics were attracting so much interest mainly because the materials on which previous hopes had fastened were just not working.

While the growth in world wind capacity has been measured in gigawatts in the last 10 or 15 years, growth in solar capacity is tallied in megawatts—a factor of a thousand smaller (see table). Total world solar capacity at the end of the last century was barely equivalent to that of a single medium-sized coal-fired power plant, and though solar capacity has grown by well over 30 percent annually since the turn of the century, it's growing from a tiny base. Almost weekly, someone somewhere in the world announces placement of a record-setting PV roof on some big public facility—a subway terminal at Coney Island in Brooklyn, a convention hall in Munich, a school or prison in California, a sanitarium in Japan. But no such installation is ever made without public subsidies, and the subsidies are huge—not the mere pennies doled out for wind-generated watts, but real money that would add up to big bucks and quickly become a troublesome political issue if solar power were to take off on a really large scale.

Just consider the situation facing taxpayers in one jurisdiction, New York State, where a farsighted and environmentally minded government has sought to jump-start a solar revolution by offering very generous incentives to any homeowner who wants to put solar panels on a roof: for a system that might provide half or two thirds of the electricity needed in a house accommodating a family of 4 or 5 people, a suitable solar array costs about $32,000. The state will reimburse half of that, $16,000, as a flat subsidy, and then another $4,000 in the form of a tax break. Yet even so, and even at low or moderate interest rates and with local electricity rates among the highest in the nation, the $12,000 the homeowner invests will not be paid back from saved electricity for more than a decade—and then only if the somewhat experimental system works as advertised, and the company providing it stays in business and honors any warranty given at the time of purchase.

Simply put, in the words of Deutch and Lester, "the economic case for photovoltaics has yet to be demonstrated."[21] That is to say, from a commercial perspective, the technology is *not* demonstrated; it is not ready to play much of a role in a concerted effort to bring down greenhouse gas emissions; and it's not yet capable of forming one of Pacala and Socolow's one-gigawatt wedges. Photovoltaics is a technology that everybody wants to believe in. But such sentiment can get in the way of coolheaded decision making, as Deutch and Lester put it, introducing their evaluations of wind and solar energy. With nuclear power, we have discovered that "it is not easy to manage technology development when there is public mistrust of the technology." With solar power, we see that "it can also be difficult to manage the development of a technology for which there is torrid public enthusiasm."[22]

Renewable energy technologies, being by nature very desirable and attractive, are readily and frequently oversold—most often by idealists, who sincerely believe they are the whole answer to all resource problems. Amory Lovins, the apostle of "soft energy paths," proposed in a famous 1976 article that the United States embark on "a path less chosen," relying on decentralized and renewable energy sources rather than centrally generated electricity.[23] Based now at his Rocky Mountain Institute near

Golden, Colorado, where he built himself a home and a headquarters around principles of conservation, efficiency, and green design, Lovins has been a steadfast and effective proponent of innovative new technology. As Smil observes, few if any have done more to promote the good cause of renewables. Yet it's also the case, notes Smil, that the share of renewables in the U.S. energy supply in the year 2000 was only a tenth what Lovins had predicted in 1976.[24] Of course Lovins is not to blame for the fact that many of his sensible ideas have not been adopted as policy. In the 1980s and 1990s, with free-market economic principles generally carrying the day, the government dropped many incentives to renewable energy development that had been put in place during the Carter administration.[25]

Just the same, the main reason renewables have failed to meet expectations is that solar energy technologies are simply not market-ready, contrary to Lovins's expectations, while wind—though much more viable—is limited by the geographical, electrical, and environmental considerations described previously. It is wishful thinking to imagine that renewables can displace more than a fraction of centrally generated electricity. Altogether, assuming wind accounts for much more power generation growth than solar, the two technologies might make in the next fifty years one or two of the one-gigawatt wedges, accounting for roughly one or two sevenths of the carbon-based energy that needs to be replaced globally.

More, perhaps much more, is to be hoped for from gains in efficiency and from conservation. As Socolow and Pacala conclude, "improvements in efficiency and conservation probably offer the greatest potential" of any means to avoid future carbon emissions, and yet "options are less tangible than those from the other categories," such as renewables, natural gas, nuclear energy, and carbon sequestration. "Improvements in energy efficiency will come from literally hundreds of innovations that range from new catalysts and chemical processes, to more efficient lighting and insulation for buildings, to the growth of the service economy and telecommuting."[26]

Making much the same point but in the context of a book that has a quite different perspective, Peter Huber and Mark Mills write that "improvements are possible whenever combustion and thermal processes are used to purify, synthesize, and treat materials. The billions of electric valves that currently control gas and liquid flows in factories and homes are being linked to sensors that monitor temperature, pressures, flow rates, and other variables. The chemical factory gets networked end to end, and dexterous robots built into the hardware take control of the boiler, the furnace, the mixing vat, and the reaction vessel. By far the most effective way to raise efficiency and to control pollution is to improve control of the flow of energy through the systems that consume it."[27] To take another specific but illuminating example, DuPont has developed a new material, Volron, which is a highly durable wire coating for use as a superior insulator in electric motors. "In DuPont's tests on electric motors," *BusinessWeek* reported, "a coating of Voltron extended the time between failures by a factor of 10, to more than 1,000 hours. And since such motors consume [a large fraction] of U.S. electric power, lengthening their life and efficiency promises big energy savings."[28]

All the possible gains from efficiency and conservation may be difficult to enumerate and visualize concretely, but this doesn't mean they're a mere matter of faith. Teams of specialists who have actually tallied up potential gains have found, if the proper policy incentives were in place, that net savings would be considerable indeed. Even the much maligned 2001 report on national energy policy prepared by President George W. Bush's vice president, Dick Cheney, acknowledged that historic gains from improved efficiency were impressive and that continued gains could be expected. Though environmentalists had expected to hear in the report nothing but "drill, drill, drill"—and many stubbornly refused to hear anything else even after it was out—the report documented that improved energy technologies over the last three decades resulted in U.S. residents paying a smaller proportion of their income for energy than they did before the 1973 oil crisis, despite a significant increase in total energy usage. Recognizing that big gains in efficiency still could be achieved, it recommended sizable tax credits for hybrid-electric vehicles and purchases of solar panels, greater use of biomass, and encouragement of sundry other renewable energy and conservation technologies. Still, as critics complained, the Cheney report basically was resigned to continued heavy reliance on domestic coal and foreign fuels.[29] The previous year, however, a report prepared by researchers at three U.S. national laboratories presented in some ways a radically different view of what the country's energy sector could look like in twenty years. Instead of an enormous and inexorable increase in reliance on fossil fuels, the report argued, coal and oil consumption can be held close to current levels.[30]

Generally speaking, savings from improved efficiency can come on both the supply and the demand side. The efficiency of all generators and fuel converters tends to improve in time, though as technologies mature, the improvements naturally come at slower rates. But there are some simple ways very large efficiencies could still be reaped in the electricity sector, if thermal heat now mostly wasted by generators were instead turned to productive use. Even the most efficient natural gas generators still give off half of the energy they consume as waste heat, which is why power plants—not just nuclear ones—typically have the cooling towers that somehow have become associated in most minds just with atomic energy. That heat, instead of being vented to the environment, can in principle be used to drive industrial processes or warm buildings. Thomas Casten and Brennan Downes, who are consultants with Primary Energy in Oak Brook, Illinois, point out that the best plants configured to provide both heat and power can achieve net efficiencies as high as 97 percent, by recycling heat and by avoiding transmission and distribution losses. If smaller plants of this kind were very widely distributed, so that their thermal energy were put to local use, Casten and Downes argue, the savings in both money and energy could be huge. Taking an extreme case—a situation in which all added energy needs through the year 2020 were met by dispersed combined heat-and-power plants—they estimate that the capital savings would come to $326 billion and the power savings to $53 billion, while carbon emissions would be cut in half.[31]

Casten and Downes may overstate the political impediments to combined heat and electricity—reformed utility regulation has done more than they acknowledge has been done to foster "cogeneration" by independent power producers[32]—and they understate the purely local impediments to siting plants in places where heat could be used effectively. Hardly anybody wants such a plant in their backyard, and increasingly, cities try to avoid having power plants within their borders at all. So it's clear that combined heat and power also is not the whole solution to the carbon problem. There is no single solution. But there is considerable potential, if only citizens can be persuaded that local siting of plants is not, as they are inclined to feel, an environmental blight but rather a boon.

One place that has showed the way, ironically, is New York City, which many Americans are predisposed to consider the worst kind of environment. It has huge power plants well within the city limits: throughout the 1960s and 1970s, for example, the Ravenswood plant on the Queens side of the East River, facing the United Nations, was the largest thermal plant in the world; it continues to produce a big fraction of the city's electricity today, burning oil most of the time. Going back many decades, waste heat from such plants has been piped underground throughout Manhattan to provide energy to big commercial and residential buildings. (That system is unique in the United States but fairly common in parts of Northern Europe.)

In recent decades, New York City has been a leader in what goes by the name of "green design" as well. That term refers partly to the greening of urban architecture in the literal sense: allowing for big urban parks, pocket parks in big complexes, and even incorporation of lawns, shrubs, and trees into actual buildings.[33] More importantly, though, green design has to do with the incorporation of environment-friendly technologies into building plans, including not just photovoltaic panels and even wind turbines but also passive solar features, ways of making buildings both retain heat and breathe air, and use of readily recyclable materials so as to minimize demands on external resources, including energy and water.[34] A number of showcase buildings have gone up in Midtown, including Condé Nast's 4 Times Square, a futuristic structure with PV panels and fuel cells providing a fraction of its tenants' electricity, and 1 Bryant Park, a couple of blocks away, across from the New York Public Library. At Battery Park City, a huge residential and commercial complex being developed on landfill adjacent to where the World Trade Center towers once stood, highly articulated green design principles have been a part of the planning process from the start. As a result, Battery Park City is a model for large new urban developments from Inchon, South Korea, to the mushrooming megalopolises of South America.[35]

From the point of view of energy and carbon mitigation, New York's single most "green" feature is without doubt the very aspect that many other Americans consider so repellent: its human density. Because of the city's compactness, there is a relative paucity of free-standing homes and of automobiles, and very high reliance on public transportation. Accordingly, New Yorkers use far less energy and emit much less greenhouse gas per capita than average Americans, but at the price of living cheek to jowl. In a *New Yorker* article that attracted attention not only from Manhattanites but also from green architects, writer David Owen introduced the apparent irony of New York's environmental correctness with this telling anecdote: "My wife and I got married right out of college, in 1978. We were young and naïve and unashamedly idealistic, and we decided to make our first home in a utopian environmentalist community in New York State. For seven years, we lived, quite contentedly, in circumstances that would strike most Americans as austere in the extreme: our living space measured just seven hundred square feet, and we didn't have a dishwasher, a garbage disposal, a lawn, or a car. We did our grocery shopping on foot, and when we needed to travel longer distances we used public transportation. Because space at home was scarce, we seldom acquired new possessions of significant size. Our electric bills worked out to about a dollar a day." That utopian community was the Upper East Side of Manhattan—in the city that most Americans think of as an "ecological nightmare," as Owen put it, but that, in comparison with the rest of the country, is a "model of environmental responsibility."[36]

In principle, greater efficiency and conservation are virtually synonymous: if you use less energy to accomplish the same thing, you use less energy. But in practice, when people are given a more energy-efficient way of doing something, they almost always do it more. As cities like Phoenix, Arizona, sprawl, citizens move into ever bigger houses with rooms that have to be heated or cooled and lawns that have to be watered, and traffic of course gets worse and worse. But as more and larger highways are built to ease traffic congestion, a side effect is what transportation planners call "induced traffic." The more good roads are built, the more people use them, the more their communities sprawl, and the worse congestion gets again.

Smil finds that this pattern is ubiquitous. Though energy use per square meter in the average American home is significantly lower than it was a generation ago, because of advances in conservation-minded design, the size of American homes has increased by more than 50 percent in the same period, so that total residential energy usage is higher than ever. Gains in the fuel efficiency of the average U.S. car since the early 1980s have been completely wiped out, and more, by the increased number of miles the average car owner drives each year. In Britain, he notes, the efficiency of street lighting has improved twentyfold since the 1920s, in terms of lumens per watt; yet because of a desire for better and more intense lighting, the amount of energy used to illuminate an average stretch of road has increased by a factor of 25 in the same period.[37] What's happened with lighting in advanced industrial countries in just the last couple of decades provides another telling example: with the invention of compact fluorescent bulbs, which are much more efficient than incandescent bulbs, consumers bought them in huge numbers; but at the same time, halogens, which are actually much less efficient than standard incandescent bulbs, also came onto the market, and as they were incorporated into all manner of cute new designs, consumers bought them too in huge numbers, more than wiping out any gains from compact fluorescents.[38] Vint Cerf, considered by those in the know to be the main inventor of the Internet (not Al Gore!), has found what's often called the "efficiency paradox" even in the digital domain: "The Internet has the funny effect of increasing the amount of travel—people using it discover places to go and people they want to meet."[39]

"Historical evidence," concludes Smil, "shows unequivocally that... advances in energy efficiency have not led to any declines of aggregate energy consumption."[40] Agreeing, Huber and Mills observe that "to reduce energy consumption, a more efficient technology has to have a greater impact in the replacement market it creates than in new markets it infiltrates."[41] But that's rarely the case. Overall, in the last two decades of the twentieth century, the energy efficiency of the U.S. economy improved by a third; yet in the same period, aggregate energy usage increased 25 percent.[42]

The paradox is not a new discovery. Fifty years ago, two students of the automotive industry, Eugene Ayres and Charles Scarlott, formulated the principle in language "that every new generation of energy pundits seems fated to repeat," as Huber and Mills put it. "Higher automotive-engine efficiencies are announced from time to time as resulting from improved engines or superior fuels," Ayres and Scarlott wrote. "But motorists have not realized any increase in mileage, since the potential efficiency increase has been offset by running more powerful engines under lighter partial loads and in hauling more tons of automobiles at higher average speed. The motorcar operator is largely to blame. He insists on excessive weight, power, and luxury. He accelerates too rapidly. His speed is excessive. He knows in a vague way, that he is paying for all this, but he feels that he is getting his money's worth of what may be called the 'amenities' of driving."[43]

From Cheney to Smil, it's agreed that prodigious gains in energy efficiency have been achieved in recent decades, and that similar gains can still be made in the decades ahead. But for those gains to be translated into actual conservation of energy and lower carbon emissions, we must discourage ourselves from using more of the more efficiently harnessed energy, either by regulatory fiat or by means of tax policy. From market-oriented conservatives like Cheney to those more inclined to social planning like Smil, analysts also generally agree that tax policy is in principle the better approach, inasmuch as higher carbon or energy taxes create a level playing field so that the most efficient means of conserving energy are chosen at the ground level, with the government "picking winners." Yet in practice that consensus breaks down, and at both ends of the political spectrum. The business-oriented conservatives are subject to pressures from industries that will be especially hurt by higher taxes, and so instead of acting resolutely on their free-market principles, they slide into a program of research subsidies and production credits—in effect giving handouts to favored industries, and indeed "picking winners"—exactly the approach they are so quick to decry when it is proposed by others with different favored clients in mind. Others, beholden to union members and pledged to defend the interests of the average citizen, are reluctant to pass tax increases that by nature are nonprogressive—that is to say, a higher emissions tax translating into higher gasoline prices will affect the industrial worker commuting to his factory every day in an old pickup or van essentially as much, in dollar terms, as it affects Bill Gates in his daily commute.

The way out of this dilemma, crudely speaking, is to make carbon or energy taxes revenue-neutral, as other countries already have done, so that the proceeds are fed back to individuals, regions, and perhaps even businesses to compensate them for the higher burdens they are carrying to bring down the country's greenhouse gas emissions. It must be borne in mind, though, that this approach is a slippery slope. Once politicians begin negotiating about just who gets what, the end result is all too often the kind of Christmas-tree energy legislation—everybody gets to put an ornament on it—that sank efforts at formulating a U.S. national energy bill in 2003–04. It would be best, then, to confine paybacks to a system of rebates in which parties adversely affected in narrowly specified ways are entitled to compensation to be used for closely construed purposes (replacement of the aging pickup by a fuel-efficient hybrid car or purchase of zero- or low-carbon generating technology, for example). Regional economic development credits would be warranted for those areas in danger of being crippled by a phaseout of coal extraction.

In light of historical experience and general technological trends, it seems evident that a tax-driven program of renewables development and green design could yield over the next twenty-five years at least two or three of Pacala and Socolow's one-gigawatt wedges, taking the world about halfway to long-term carbon stabilization. But renewables and green design will not by themselves accomplish the whole job of carbon stabilization. This is partly because the automotive sector is so uniquely resistant to energy conservation, and partly because of another conundrum: if the answer in that area is the adoption of plug-in hybrid-electric cars, which promise much greater fuel efficiency, demand for electricity will increase substantially as such cars are adopted; if the answer is hydrogen, most of it will likely have to come from natural gas, at least in the short run. But that means additional pressure on natural gas prices, which already have been climbing steeply, and less natural gas to substitute for coal-generated electricity.

CHAPTER 10

___

Natural Gas, Gasoline, and the Vision of a Hydrogen Economy

O NE OF THE more profound ideas about energy to have emerged in the last generation or so is this: the march of material progress, from the first hunt-and-forage cultures to today's most advanced postindustrial societies, is in essence a triumphal procession from high-carbon, low-hydrogen fuels like wood to low- or zero-carbon, hydrogen-intense fuels like natural gas. The notion is that as material civilization has advanced, we have moved steadily from less efficient, dirtier fuels to more efficient, cleaner ones. The hydrogen-intense fuels are better because when hydrocarbons are burned, a much larger fraction of the energy gain comes from the oxidation of the hydrogen than from the conversion of carbon to carbon dioxide. The ultimate stage of civilization, in this view, which is most closely associated with Professor Jesse Ausubel of Rockefeller University in New York City, is reached when pretty much everything runs on pure hydrogen—the so-called hydrogen economy.

In 1996, the American Academy of Arts and Sciences invited Ausubel to produce a special issue of its journal, *Daedalus,* with his basic ideas about hydrogen and carbon in mind. Ausubel gave the issue an arresting title, "Liberation of the Environment." In his introduction, he spelled out the logic, noting that per atom oxidized, hydrogen yields about four times as much energy as carbon. Accordingly, as humans have mastered nature, freeing the environment from the curses of pollution and resource depletion in the process, they have moved from primitive carbon-rich fuels like sticks and dung to today's advanced fuel of choice, natural gas. "Wood weighs in heavily [with] ten effective [carbon atoms] for each [hydrogen atom]. Coal approaches parity with one or two Cs per H, while oil improves to two H per C, and a molecule of natural gas (methane) is a carbon-trim CH_4."[1] Hydrogen, obviously, is pure H.

It would seem, from the way events unfolded in President George W. Bush's first term, that somebody must have whispered something in his ear about Ausubel and his ideas. In January 2003, two months before Bush launched the Iraq war, he put his administration squarely behind a concerted effort to convert the U.S. automotive sector to hydrogen. The general idea is to eventually have all motor vehicles powered by fuel cells—devices that produce energy by oxidizing pure hydrogen, yielding only water as a by-product—and to obtain the hydrogen fuel by electrolyzing water, converting natural gas, or obtaining it from advanced clean-coal technologies being developed with government support.

This notion of a hydrogen economy, a natural derivative of Professor Ausubel's worldview, was enormously compelling as a vision of what the world will eventually look like—not just to Bush's devoted supporters but also to some clean-energy boosters like Amory Lovins. But as a guide to making energy policy for the years immediately ahead, it suffered from a certain absence of common sense about what could reasonably be achieved, and when. Almost everybody could agree that ultimately the world will be powered mainly by hydrogen (with the important qualification that zero-carbon technologies like fission and fusion technology also will play important roles). But that's not to say that the hydrogen vision will be realized in time to help with any of the problems immediately at hand, be they dependence on imported oil or escalating emissions from fossil fuel combustion. Meanwhile, technologies and polities that can help cut fuel dependence and emissions are neglected.

Almost the moment the hydrogen initiative was announced, it came under such thoroughgoing and scathing criticism that most of the salient points have become clichés even to those following the debate only cursorily. Still, for the record, they must be recounted. In the first place, hydrogen is not a fuel like wood, coal, oil, or gas that can be simply carried anyplace and burned, using simple and well-established technologies. It's an energy carrier, much more like electricity, that must be generated from other sources and then distributed via an elaborate infrastructure, which has to be versatile and reliable enough to accommodate needs that will vary quite drastically in the course of the day and night, and in most places, from season to season. In other words, the infrastructure has to be big, highly redundant, and very expensive.

The second point follows directly from the first. Since hydrogen has to be generated, it is only as pure as the technologies that make it. If, for example, the hydrogen is extracted from water by means of electrolysis, enormous amounts of electricity are required, and that electricity right now is made from dirty coal. If it were made instead from natural gas, which is the most plausible technique at present, or from coal, which could become a plausible technique in the not too distant future, then the hydrogen is pure only if the carbon emitted as a by-product is captured and sequestered. And that, as discussed in chapter 8, is not as yet a commercially demonstrated technology.

Third, the technology used to convert hydrogen to usable energy, the fuel cell, though well proved in principle, is not nearly as market-ready as expected ten years ago or even—to be fair to Bush—when he unveiled his initiative in January 2003. Invented in the mid-1800s, the fuel cell was developed into a real-world working device a century later in the context of the U.S. space program, as a way to generate electricity from hydrogen in the Gemini and Apollo vehicles. That led to the highly appealing notion that similar devices, if made cheaper and more efficient, also could be used to power vehicles on earth, as well as homes, institutions, and perhaps even factories. The Connecticut company that developed the fuel cells for the space program has sold several hundred commercial derivatives to institutions like hospitals, where they are capable of supplying power needs independently of the sometimes unreliable electricity grid.[2] Another company in the same state, developing an alternative approach, deployed a large battery of its cells in a pilot plant in Santa Clara, California, to show that in principle fuel cells can even serve as a source of centrally generated electricity.[3]

As concerns about diminishing oil supplies became more acute in the 1990s and alarm rose about climate change, fuel cells came to be widely seen—and not incorrectly—as the most promising way of ultimately breaking dependence on the liquid fuels that power transportation everywhere. By the beginning of this century, every major automobile company in the world had an ambitious fuel-cell development program, and some were beginning to put experimental vehicles—usually fleet vehicles like delivery vans or buses—on the streets. DaimlerChrysler, probably the world leader in this endeavor, teamed up with the Canadian company

Ballard, a highly regarded fuel-cell startup in Vancouver, to develop and test vehicles from Stuttgart to Beijing. Meanwhile, the exciting notion developed that use of fuel cells in vehicles and homes could be complementary and mutually reinforcing. Plug Power, a startup in Latham, New York, envisioned a fuel cell-powered car that could be driven all day, then plugged into the home at night to power the house; alternatively, a home hydrogen-generating system could power the house and the excess fuel could run the car. Promising that it had a technology almost ready, as the company name implied, to plug right into the house or car, Plug Power entered into a distribution agreement with General Electric in the late 1990s, promising that it would have tens of thousands of cells ready for GE to sell by 2001.

When the time came, however, Plug Power was unable to deliver, resulting in punitive action by the energy giant and a class action suit brought by shareholders, who complained that the startup had misrepresented its prospects. Plug Power survived by the skin of its teeth, but its stock has taken a beating in the intervening years, as has Ballard's—a disconcerting development that did not go unnoticed by advocates of clean, alternative energy sources.[4]

Around the time Bush launched his hydrogen initiative, at the 2003 Detroit auto show, the chief executive of Nissan told a reporter for *The New York Times* that a fuel cell-powered vehicle might cost $700,000. Two years later, the engineer heading up a General Motors fuel cell project, the Sequel, told the same reporter that fuel cell cars were still at least ten times too expensive to be commercially feasible.[5] By then it was apparent to all dispassionate observers that fuel cells were far from ready to meet the technical specifications required for most uses in the automotive sector, and even farther from being commercially mature. The two companies most frequently mentioned as leaders of the effort to develop commercially viable fuel cells—Ballard and Plug Power—were both in trouble. And that was just the beginning of the problems with the hydrogen vision.

Storing the hydrogen needed to run a fuel cell onboard a vehicle is perhaps the biggest problem. One possibility, in principle, is to carry it in the form of natural gas and extract hydrogen for the fuel cell as needed. But that approach evidently is too cumbersome and now is rarely discussed. The alternatives are to store the hydrogen as a liquid, a highly compressed gas, or bound in certain chemical structures that enable it to be easily released in two-way reactions. Since hydrogen is kept liquid only at temperatures close to absolute zero, requiring expensive and complex refrigeration systems, this is not practical. The only immediately practicable approach, in fact, is to store the hydrogen under very high pressure in containers made from the strongest and lightest known materials. For such containers to win wide acceptance among consumers, car buyers will have to be convinced that the canisters are no more likely to explode in an accident than a gasoline tank, and that the hydrogen will not escape and catch fire. At present, no container exists that is capable of carrying enough hydrogen to give a car an acceptable range—generally taken to be at least a couple of hundred miles. High-pressure hydrogen containers have been developed over several decades for use in the aerospace industry, to very demanding specifications. They have gotten steadily better, but they are not nearly good enough for wide vehicular use as yet, and there is no guarantee that one good enough for automobiles will be devised soon—or ever.

Long term, the smart money has been betting on chemical storage. Robert Stempel, a famous former automobile industry executive, has taken the helm of a Michigan-based company called Energy Conversion Devices, which is closely allied with ChevronTexaco Ovonic Hydrogen Systems, also in Michigan. "ChevronTexaco," together with Stempel's involvement, tells you that the biggest of big companies are seriously interested in this technology. The word "Ovonic" in ChevronTexaco Ovonic derives from the surname of Stanford Ovshinsky, a brilliant inventor with a number of breakthrough or near-breakthrough technologies to his name, including the ubiquitous nickel-metal hydride battery.[6] The general idea of chemical storage is to store the hydrogen in solid form in materials called metal or chemical hydrides, which can be induced to take up or release hydrogen, absorbing or giving off heat. Though the solid storage approach has a lot of support, it too will not be commercially realizable without fundamental scientific and technological advances.

Even if and when all the basic technical problems of storing and converting hydrogen are simultaneously solved, there will remain the gargantuan task of producing hydrogen and getting it to widely dispersed points of use. One intriguing possibility is to convert it from natural gas at the home, for both vehicular and home use, relying on the regular gas feed as the source. That approach has been floated like a trial balloon by Plug Power but has yet to be commercially demonstrated. Generally it's assumed that hydrogen will have to be produced on a very large scale at central facilities, perhaps first from natural gas and later from coal, and then distributed to filling stations either on large refrigerated trucks (which is done now) or by means of a pipeline system like the ones currently distributing natural gas all over the United States and Europe. Scattered hydrogen filling stations have been established in a number of countries, to test and demonstrate the viability of fueling cars essentially the same way gasoline and diesel vehicles are currently fueled. But only tiny Iceland, with the means of generating prodigious amounts of hydrogen locally using geothermal energy, is in a position to contemplate basing its entire transportation system on hydrogen anytime soon. Everywhere else, tentative trials are confined to fleet vehicles that can be fueled at central dispatch points.

With so many fundamental problems requiring simultaneous and mutually consistent solutions, it's hardly surprising that the hydrogen vision has fared poorly with just about every group and organization that has taken an unbiased look at it. In March 2004, the public affairs panel of the American Physical Society—the professional organization that represents most U.S. physicists—issued a report that delivered a scalding verdict on the president's notion that hydrogen vehicles can be commercially competitive by 2020. The panel concluded that no current technologies represent competitive options. "The most promising hydrogen-engine technologies require factors of 10 to 100 improvements in cost or performance in order to be competitive.... Current hydrogen production methods are four times more expensive than gasoline. Finally, no material exists to construct a hydrogen fuel tank that meets the consumer benchmarks.... These are enormous performance gaps. Incremental improvements to existing technologies are not sufficient to close all the gaps. For the hydrogen initiative to succeed, major scientific breakthroughs are needed."[7]

The month before that, a panel of the National Academies of Engineering produced a report with essentially similar conclusions: for vehicular use, current fuel cell lifetimes are "much too short" and costs are at least ten times too high to be competitive; while the costs of building a hydrogen distribution system could be spread over a great many users, "the transition is difficult to imagine in detail"; costs of distributed hydrogen production must be cut "sharply"; producing hydrogen using electricity made from renewable resources would require "breakthroughs" to be competitive; if hydrogen were extracted from coal, "massive amounts of CO_2 would have to be captured and safely and reliably sequestered for hundreds of years." Thus, "although a transition to hydrogen could greatly transform the U.S. energy system in the long run, the impacts on oil imports and CO_2 emission are likely to be minor during the next 25 years."[8]

A really fundamental problem, enunciated by the chief executive of Shell Hydrogen, Donald Huberts, is that the energy gleaned from hydrogen will always be more expensive than the energy contained in the sources used to generate it, because of the extra processing involved. "It will be competitive only by its other benefits: cleaner air, lower greenhouse gases, etc.," Huberts told a reporter for *Science* magazine.[9] Making the same point—that hydrogen always will take more energy to produce than it will yield—Professor Franco Battaglia of the University of Rome put it like this: "You can buy an apple for one euro. If you really want an apple, you might pay five euros. You could even pay a thousand euros. But you would never pay two apples."[10]

Accordingly, people at both ends of the political spectrum have been asking why technologies that will not be of much help for at least a couple of decades get so much attention, when urgent problems cry out for solutions right now. Introducing a special report on hydrogen in the August 2004 issue of *Science*, the magazine's editor in chief, Donald Kennedy, observed, "The trouble with the plan to focus on research and the future, of course, is that the exploding trajectory of greenhouse gas emissions won't take time off while we are all waiting for the hydrogen economy."[11] Around the same time, a coalition of national-security hawks and clean-energy advocates issued a report, "Set America Free," in which they bemoaned the country's ever greater dependence on unreliable and hostile oil suppliers, its focus on solutions that do not address that problem, and its relative inattention to technologies—notably hybrid-electric and alternative-fuel vehicles—that could reduce U.S. vulnerability immediately. Original signatories of the report included Frank Gaffney, a top security official in the Reagan and senior Bush administrations; James Woolsey, the director of the Central Intelligence Agency in President Clinton's second term; and Robert McFarlane, Ronald Reagan's national security adviser.[12] When the report was sent to President Bush in the form of an open letter in March 2005, Timothy Wirth, a former Colorado congressman who served in the Clinton administration as the State

Department's head of global environmental affairs, and John Podesta, Clinton's chief of staff, signed on as well.

The hydrogen economy still has fans, on the political left as well as the political right, notably Lovins and Jeremy Rifkin, who are often quick to latch onto what seem to be exciting new trends.[13] Bush's vision naturally met with an especially enthusiastic reception in the U.S. automobile industry, which liked the idea of getting research subsidies for advanced vehicle development without having to actually meet stricter fuel-economy standards right now. Taking its cues from Bush's FreedomCar initiative, General Motors organized a traveling road show to demonstrate concept hydrogen cars like Hy-Wire and HydroGen 3, which it invited press and dignitaries to try out in locations around the country.[14] But most disinterested, sensible people wondered—with Gaffney and McFarlane, Wirth and Podesta—why we should go all out for a whole loaf that may never actually materialize when a good half loaf is already available.

From 2003 to 2005, when the United States was supposed to be setting off on the road to the hydrogen economy, a funny thing happened. A technology that was generally seen as a minor detour or way station, the hybrid-electric car, took off like a rocket. When Toyota offered the Prius, in alternative versions in Asia, Europe, and the United States, consumers were invited to spend at least several thousand dollars more for a vehicle that would offer only modest improvements in fuel economy compared to other automobiles of standard design. Just as good or better fuel efficiencies could be obtained at lower prices from regular economy-size cars, and, of course, the long-term performance of novel vehicles like the Prius was uncertain. So one might have assumed that only enthusiastic environmentalists and car hobbyists would buy the hybrids. Instead, in the 10 months from September 2003 to July 2004, the Prius was the fastest-selling car in the U.S. market, which prompted Toyota to double its production in 2005, to 15,000 units per month.[15]

Gains in fuel efficiency are relatively modest in hybrid-electric cars like the Prius because they ultimately get all their energy from their gasoline fuel. They carry a bigger but essentially conventional battery pack that, combined with electric motors and a complex electronic control system, enables the car to be driven electric-only at low speeds for limited distances, in the rpm range in which electric propulsion is most efficient. But most of the time the car is being powered by its internal combustion engine, and all the battery charging is done by that engine—such cars cannot be plugged into a wall outlet to be recharged. However, if they were equipped with batteries capable of storing much more energy—scaled-up versions of the lithium-ion batteries found in many handheld devices, for example—they could yield gains in fuel efficiency and cleanliness that would be truly impressive. A Prius produces about a quarter less carbon emissions and 15 percent less pollutants than a comparable mid-size car of standard design. But if the hybrid were turned into a plug-in, the total amount of energy it uses and the quantities of carbon and pollutants it emits might be cut by half or even more.[16]

The major manufacturers of the hybrids on the market today—Toyota, Honda, and Ford—have been reluctant to introduce plug-ins because they had assumed consumers would not be willing to pay yet another price premium, perhaps $5,000 or $10,000 extra. But with oil prices rising sharply in 2003 and 2004, U.S. consumers proved remarkably willing to spend more money to economize somewhat on fuel and at the same time do something positive for the environment. Meanwhile, some clean-car promoters and automotive tinkerers eagerly installed enhanced battery packs in cars like the Prius to show that they could be easily converted to hybrids. One such Prius converter, Ron Gremban of CalCars (the California Cars Intitiative, in Palo Alto), conceded that the modified car did not perform as well as the original and cost considerably more. But "a company with the resources of a Toyota, Honda or General Motors could build a more elegant full-function version for far less money," he argued.[17]

By 2005, DaimlerChrysler was getting set to start road-testing in the United States a plug-in hybrid-electric delivery van called the Sprinter. If all went according to plan, a wider trial, organized with assistance from the Electric Power Research Institute in Palo Alto, would take place in 2006. By then, a fairly wide range of regular hybrid-electric cars (without the plug-in capability) would be available not only from Toyota, Honda, and Ford but also from General Motors, Nissan, and DaimlerChrysler. And many more companies would be seriously pursuing commercialization of plug-in variants. For by now most of the major automakers have reorganized their R & D programs to accelerate development of hybrid-electric cars and downplay hydrogen-powered fuel cell cars. If oil prices stayed high or went even higher, or if governments imposed more stringent fuel-efficiency standards or higher gasoline taxes, hybrids clearly were going to be the wave of the future.

An additional way of reducing dependence on foreign oil imports while at the same time cutting greenhouse gas emissions, also highlighted in the Gaffney manifesto, would be to sharply boost production of liquid fuels from crops. In principle, the potential may be very great, but in practice there are limits. Pacala and Socolow estimate that to make one of their seven "stabilization wedges" needed to keep total global carbon dioxide emissions at present levels, the United States or Brazil would have to increase ethanol production a hundredfold. That would entail using a sixth of the world's cropland for ethanol production. There is also some question about the net energy gain from ethanol use, as growing the crops is usually very energy-intensive. Ultimately, however, the most serious issue of all is how much drivers are willing to spend for greater energy independence and less risk of climate catastrophe.

During the two years following the U.S. decision to go to war in Iraq, world oil and U.S. gasoline prices increased about one third. Together with mounting concern about rising greenhouse gases and local pollution, the higher energy costs gave consumers and regulatory authorities powerful incentives to adopt more fuel-efficient and cleaner automotive technology. These fundamental forces delivered a big boost to not only the hybrid-electric car but also other alternative-fuel vehicles and more fuel-efficient automotive technology generally. Those price increases may be temporary, however, and if oil prices drop again, efforts to make vehicles more fuel-efficient will be undermined.

But for the weak U.S. fuel-efficiency standards, and especially the loophole that has allowed sport utility vehicles (SUVs) to be classified as light trucks, the overall fuel efficiency of U.S. vehicles could easily be twice what it is today.[18] From a purely technical point of view, there's almost no limit to the fuel savings that can be achieved in the automotive sector. Given enough incentives to purchase better vehicles, through higher fuel taxes or stricter fuel-economy standards, gasoline consumption and emissions could be reduced by half almost overnight. What actually is accomplished comes down, basically, to what prices and rules the public can be persuaded to accept or impose on itself.

That of course is the rub. While many experts on the oil industry, including members of the Bush administration and people close to the administration, believe that the world has reached a point where oil consumption has begun to permanently outstrip additions to reserves, so that prices will remain high from here on out, there is no guarantee that will be the case. If the notion of "peak production" holds, then the problem of achieving much better fuel economies and lower emissions in the automotive sector may take care of itself: at least in the United States, the higher gasoline prices might by themselves be an adequate incentive to develop and adopt new technology. (In Europe and Asia, where gasoline is taxed much more heavily to begin with, the higher oil prices have had much less impact on consumer behavior or policy. Eager to economize on gasoline, more than half the drivers in continental Europe now own diesel cars, with very unfortunate effects on local and regional pollution.) But what if, contrary to the pessimists, it proves possible to locate and exploit new oil reserves, and gasoline prices drop? In that case, dramatic automotive fuel savings would be obtained in the United States only if taxes were imposed to deliberately drive prices higher.

Studies of what economists call the price elasticity of gasoline demand—the amount that demand for energy increases or decreases with a given change in price—usually indicate that the long-term value is roughly .5. That is, if energy prices increase 100 percent, then eventually consumption of gasoline will drop by 50 percent.[19] Thus, if Americans were willing to let their gasoline price go from roughly $2 to $3 per gallon in mid-2005 to about $5 to $6, then by the end of this decade or the middle of the next, national gasoline consumption—and greenhouse gas emissions from the vehicle sector—could be about half what they are today. This would amount to having the United States do its part to achieve one of the seven "wedges" required to stabilize global carbon dioxide levels at double their preindustrial value. With the automotive sector accounting for about a third of U.S. greenhouse gas emissions, if automotive fuel efficiency were improved by half and automotive emissions reduced by half, total greenhouse gas emissions would be 15–20 percent lower—close to what's needed to get into step with the Kyoto program.

But getting the U.S. driving public to accept gasoline prices that are twice as high as today's—whether by allowing them to rise naturally from market forces or by means of taxes—is a tall order. So too is trying to accomplish the same objectives via much stricter fuel economy standards. But that is not the only serious challenge. Even with much higher gasoline prices or a much stricter regulatory regime, to the extent fuel economies were achieved by deploying plug-in hybrids, the reduced gasoline consumption would come at the price of higher electricity usage. That would help make the United States more independent of oil producers, but it would not help much with greenhouse gas emissions—it would just shift the problem of how to reduce emissions from the automotive sector back to the electric sector, where even more electricity would have to be generated, in the absence of other incentives or disincentives, from dirty, carbon-intense coal.

To the extent gasoline savings are achieved through direct use of compressed natural gas or, in the much longer run, through hydrogen derived from natural gas, the potential savings in greenhouse gas emissions are more promising, though even they have been overstated. Joseph Romm, a top energy official in the Clinton administration, has argued that natural gas contributes more to reduced greenhouse gas emissions if it is burned directly to generate electricity than if it is used,

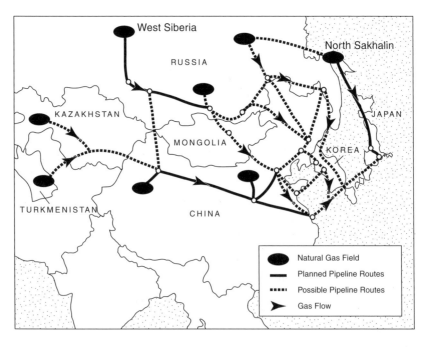

The Global Scramble for Natural Gas: Major Asian Pipeline Projects
Source: FSI Energy

instead, as a fuel source for hydrogen-powered vehicles.[20] In terms of energy independence, the gains from greater use of gas are slim or even nonexistent. Although on a global scale natural gas is thought to be enormously abundant, indeed almost unlimited from a practical point of view, the gas is not always where it's needed most, and everybody wants it. The United States is not able to meet all of its own needs even at current levels of consumption and projected levels of growth. This is why there is pressure for construction of a new pipeline to bring gas from Alaska down to the Lower 48, and why the former Federal Reserve chairman Alan Greenspan called for construction of more port terminals to receive liquefied natural gas from abroad.

The world's most prodigious natural gas reserves are in Russia and central Asia, and increasingly, there is acute competition between China and India and between Europe and the United States for access.[21] The Europeans and Asians have better access to Central Asian gas than the United States does, and arguably they need it much more. The very rapidly growing economies of China and India will take a large and growing share (see map), and Europe probably will as well. Accordingly, the United States should assume that natural gas will not be available at an attractive price to meet every conceivable need.

Natural gas, besides being a clean and relatively inexpensive fuel, is a source for chemicals widely used in industry. As a fuel, it has replaced coal and oil ubiquitously in residential and commercial heating, and during the 1990s it also was the fuel of choice for new central electricity generators all over the world. If it continues to be substituted for coal and oil in the next two decades at the rates seen in the last two decades, this would contribute greatly to reducing greenhouse gas emissions. But natural gas cannot do the whole job, and probably not more than a modest fraction of it.

CHAPTER 11

A Second Look at Nuclear Energy

IF THE WORLD'S greenhouse gas emissions are to be kept from more than doubling in this century, the United States is going to have to do not just its fair share, so to speak, but a little more. This is because as the world's richest and most highly endowed country, it can afford to do more; because U.S. use of energy is singularly extravagant; and because the world's poor countries cannot be stopped from developing as fast as they can and therefore using more energy than at present.

An affordable and achievable energy future for the United States would look roughly like this: in the next decades, economic growth will be achieved without any net increases in energy demand; conservation and efficiency, encouraged by some combination of carbon emissions taxes and gasoline taxes, or by a cap-and-trade system for carbon and fuel-efficiency requirements for automobiles, will take care of that. At the same time, combustion of coal by conventional means will be phased out: until advanced coal gasification and carbon sequestration technologies are feasible on a large scale, carbon emissions from the coal sector will be reduced by switching to low-carbon and zero-carbon fuels. Wind but not solar energy will be able to provide a sizable fraction of the electricity now provided by coal, and natural gas ought to be able to take another bite, if supply problems are solved. (As noted previously, at least one major pipeline will be needed to carry gas from Alaska to the Lower 48, and a network of port terminals to handle seaborne carriers of liquefied natural gas.) But with natural gas already in high demand and short supply, even without the added pressures on natural gas availability that will result as the automotive economy ultimately switches over to vehicles powered by fuel cells, natural gas cannot be expected to get the whole job done. Accordingly, it will be desirable if nuclear power—an even better substitute for coal than natural gas—can play a part as well.

Consider, again, the basic numbers. Per unit of energy consumed, electricity generated from natural gas produces about half the carbon emissions of electricity generated from coal. Electricity generated in nuclear reactors produces, for all practical purposes, no carbon emissions. Accordingly, with coal combustion in the U.S. electricity sector accounting for roughly a third of U.S. greenhouse gas emissions, if (to take the extreme case) all that coal-fired electricity were replaced by nuclear electricity, the immediate effect would be to reduce U.S. carbon emissions by about a third. If, alternatively, it were all replaced by natural gas, the reduction would be perhaps a fifth or a sixth.

Looked at in this perspective, it is tempting to say that the United States simply cannot get along without increased reliance on nuclear energy. Having long refused to join in the Kyoto effort, and having allowed its greenhouse gas emissions to steadily rise, it now—once it decides to get with the program, that is—will have to cut emissions much more radically than if it had joined in the late 1990s, when it originally intended to. At that time, meeting Kyoto targets would have required cuts of only about 7 percent. Now, 10 years later, the cuts will have to be closer to 20 or 25 percent. And to go beyond that and do better than Kyoto, which is what the situation requires, will call for cuts on the order of a third.

It's hard to see how that goal could possibly be achieved without greater reliance on nuclear energy. But of course it's no good to say we have to do something, even if it has undesirable effects, because we just can't get along without it. If something's just plain bad, you don't do it no matter how much you may feel you need to. But is nuclear energy bad? Should we forego it on principle? Or is it, on balance, a positive good?

The first and most important thing that can be said of nuclear reactors is that they work. In contrast to many of the other energy technologies that have been reviewed as candidates for supplanting carbon-based fuels—whether it's central generation of electricity by very large solar arrays, carbon sequestration, or vehicles powered by fuel cells running on hydrogen—nuclear power is a well-proven technology capable of producing electricity at costs that are commercially competitive right now.

It was not always so. Back in the late 1970s and early 1980s, when a lively debate over the nuclear option erupted in the United States, Japan, and above all in Europe, the performance of nuclear reactors was often poor. Reactors were enormously expensive and complicated to build, and were constantly in need of special maintenance or repair, which was devastating to the case for nuclear energy. Since reactors cost much more up front than any of the other major sources of electricity, their economic viability depends on lower operating costs. If the reactors are not actually operating much of the time, then it obviously was a mistake to deploy them in the first place. In fact, in the late 1970s and early 1980s, many of the world's reactors were running barely more than half the time. This—not the wide popular opposition to nuclear energy—is the main reason utilities worldwide stopped ordering any new reactors in the last two decades of the twentieth century. From roughly 1980 to 2005, not a single order was placed in the United States, and hardly any were placed elsewhere.

During those decades, however, reactor performance began to steadily improve, without attracting too much notice from the general public. The numbers are impressive: in 1980, U.S. nuclear reactors were generating electricity only 56.3 percent of the time on average; in 2004, the reactors were running 90.5 percent of the time (see graph). During this period, the anticipated life expectancy of reactors also increased notably—an important consideration, given their very high capital costs (though perhaps not quite as important as one might suppose[1]). In the early years, many nuclear critics believed that the materials in the power plants might degrade faster than nuclear proponents claimed, mainly because of intense bombardment by radiation. But it turns out that most reactors are holding up better than expected. In the United States and in some of the other countries that went nuclear early on, utilities that had obtained licenses to run nuclear plants for 40 years are now applying to extend those licenses for another 20 years.

In several large advanced industrial countries, nuclear energy has been supplying very large fractions of total electricity for decades, without any major mishaps. In France, which has probably the best reputation for strong nuclear management, reactors supply three quarters of the country's electricity. In Sweden and South Korea the proportion is roughly 50 percent, and in countries like Japan, Germany, and the United States it hovers around 20 to 25 percent. Even at the low end of that range, 20 percent represents a huge quantity of energy—only coal

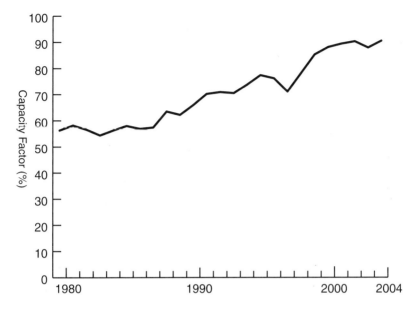

Dramatically Improved U.S. Nuclear Power Plant Performance
Source: Nuclear Regulatory Commission

produces much more in the United States, with natural gas running about even with nuclear.

Despite those numbers, nuclear construction has remained at a standstill almost everywhere except in East Asia and India; furthermore, some countries have decided in principle to phase out all reliance on nuclear reactors as fast as possible. Austria already has terminated its nuclear program, and Sweden and Germany are more or less determined to do so. The global implications of such commitments need to be kept clearly in view. To the extent countries decide to eliminate nuclear energy, all the more low-carbon or zero-carbon energy must be found to replace it. The Germans believe they can pull this off, mainly by continuing to deploy wind turbines; the Swedes cling to the notion that they can replace nuclear electricity with electricity generated by combustion of carbon-neutral biomass, and perhaps they too can succeed, though successive governments have dragged their heels.[2] Conceivably the United States could effect a nuclear phaseout too, but the scope of its problem is even greater than in a country like Germany, because of its tardiness in reducing greenhouse gas emissions. Replacing the electricity produced from coal with low-carbon or zero-carbon energy means finding new ways to generate about half the electricity now consumed in the United States. If nuclear power were phased out too, nearly three quarters of the country's electricity would have to be produced some other way.

Why did countries like Sweden and Germany decide to reject nuclear energy after building a sizable nuclear capacity? The impact of the Chernobyl catastrophe, in April 1986, was of decisive importance. What happened at Chernobyl was, in the peculiar jargon of the trade, a "worse than worst-case accident." A string of very improbable or seemingly very improbable circumstances conspired to make the reactor actually blow up, almost like a bomb, and send a plume of highly toxic radiation into the atmosphere. This laid waste to vast areas of the Ukraine and Belarus, requiring the evacuation and permanent resettlement of hundreds of thousands of people. For decades, representatives of the nuclear industry had told the public that the association of nuclear energy with the mushroom cloud was erroneous—that reactors could not explode, or at least that a nuclear explosion in a reactor was flatly impossible. Chernobyl showed definitively that this was wrong, and the impact on public opinion was enormous. People could tell that self-described experts had not been altogether truthful, or that they simply hadn't known what they were talking about.

Even in the immediate aftermath of the disaster, reactor specialists in the West had difficulty grasping that it was something more than just an aggravated meltdown—the kind of accident generally dubbed the "China syndrome." The conventional view, initially, was that there was a loss of coolant, which caused fuel to melt, and that the melting fuel in turn had caused water to convert explosively to steam, so that the reactor containment burst. That would have been bad enough, but what actually happened was even worse.[3] Because of a singular defect in the Chernobyl reactor's design—it was one of a large class of Russian reactors dubbed the RBMK—at low power levels, counterintuitively, if there was a loss of coolant, reactivity, instead of dropping, would rise sharply, so that an insidious feedback mechanism would set in. As water was lost, reactivity would rise, causing more water to convert to steam, decreasing the water density further, so that reactivity would rise still more, and so on.

The night of April 25–26, 1986, as operators working a night shift initiated a very ill-advised experiment, they inadvertently triggered the insidious feedback loop, so that in seconds, the reactor went "super-promptcritical." In plain terms, a self-sustaining, self-escalating nuclear chain reaction began that could not be stopped by any physical means once it crossed a certain threshold. The power level would just keep going exponentially higher until the reactor containment burst, causing the reacting fuel to disperse—that is to say, explode. The runaway reactivity caused violent steam explosions at Chernobyl—almost certainly two. But the underlying cause was that the nuclear fuel itself had begun to explode.[4] Accordingly, the most honest and credible Russian experts on nuclear technology concluded correctly that the accident was a "nuclear explosion," though not nearly as fast or violent an explosion as the kind induced by an atomic bomb. To distinguish what happened at Chernobyl from an atomic blast, Christopher Flavin, the president of the Worldwatch Institute in Washington, D.C., aptly called the event a "slow nuclear explosion."[5]

The precipitating cause of the accident was that the reactor's power suddenly rose sharply when its operators were shutting it down, to see if residual energy in the turbines could be used to run its controls in the event of a loss of electricity to the plant from the grid. As cooling water flashed to steam, the pressure pulse lifted the head of the reactor, the lid where all the fuel and control rods penetrate the block—a particularly vulnerable part of a reactor. (The reactor's head also was poorly designed, and surprisingly little extra pressure would suffice to separate it from the reactor, snapping all the rods.) After that initial, relatively small explosion, all the remaining water in the reactor flashed to steam and escaped, triggering the feedback mechanism once again, so that now there was a second and much more violent explosion. That was the one that sent a huge plume of radiation into the atmosphere, which was soon detected as far away as Scandinavia and ultimately around the world.[6]

It's not hard to see why, in the aftermath of Chernobyl, citizens everywhere decided it was time for a nuclear moratorium, or at least a radical slowdown. Obviously the industry had not leveled with the public about all the dangers associated with nuclear power, and that brutal truth needed to be digested. What was widely overlooked, however, and continues to be generally overlooked to this day, is that the kind of nuclear explosion seen at Chernobyl was unique to the RBMK reactor. The so-called light-water reactors[7] built almost everywhere else in the world cannot experience the insidious feedback effects that caused the Chernobyl reactor to blow up, and in fact these reactors—whether they are the pressurized water model or the boiling water variant—have a very important inherent safety feature. Because this kind of reactor depends on the presence of water for the nuclear reactions to sustain themselves, if water is lost, the reactor shuts down automatically. There remains the very serious danger that residual heat in the nuclear fuel will melt through the reactor containment, unless cooling water is very promptly restored. But the nuclear fuel in a light-water reactor cannot actually explode the way the fuel in the Chernobyl plant did.

Of course, the risks associated with a complete reactor meltdown, the China syndrome, are not to be minimized. The reactor at Three Mile Island near Harrisburg, Pennsylvania, came alarmingly close to a complete meltdown in 1979; if the melting fuel had penetrated the bottom of the containment, huge amounts of radiation would have gone into the Susquehanna River, which flows into the Chesapeake Bay, one of the world's great estuaries. That alone would have made Three Mile Island an unparalleled environmental catastrophe. The regulatory response to TMI, however, was vigorous. If the poorly digested lesson of Chernobyl was that all RBMKs should be decommissioned as fast as humanly possible, the lesson of TMI was that the safety systems in light-water reactors needed to be greatly strengthened and that ongoing scrutiny of reactor operations needed to be much sharper. After elaborate safety reviews, the U.S. Nuclear Regulatory Commission imposed very costly upgrades to existing reactors, requiring much more redundancy in crucial safety systems. And several years ago, when an aging reactor in Ohio was found to be suffering a significant amount of previously undetected corrosion in the critical reactor head, the NRC acted emphatically, requiring all similar reactors to be promptly shut down for special inspections. The Japanese and Canadian nuclear authorities ordered similar fleetwide reactor shutdowns in recent years, when systemic safety lapses came to light.

The latest authoritative international report on the consequences of Chernobyl estimates that 4,000 deaths will ultimately be attributable to the accident.[8] There were about 2,000 additional cases of thyroid cancer caused by absorption of radioactive iodine; 9 ended in death. Fifty people died from exposure to very large quantities of radiation immediately after the accident. And, with 5 million people still living in contaminated zones 20 years after the accident, the cumulative economic cost of the disaster is virtually incalculable.

It is an experience that should not and need not ever be repeated. The moral of the tragic story is that no more Chernobyl-type reactors should be built, and that those operating in Russia and the former Soviet states should be shut down as soon as possible.

The net effect of much more vigilant regulation has been to help improve nuclear performance but also substantially boost the price of nuclear electricity. The whole subject of how the cost of nuclear-generated electricity compares to the cost of electricity generated by other means is enormously complicated and mired in controversy, partly because virtually all significant energy sources are heavily subsidized in one way or another, and partly because what economists call "externalities"—the social and environmental burdens associated with producing and using a fuel that ought to be incorporated into its monetary cost—are almost never reflected in real-world prices. Probably no two people agree on what subsidies and social costs should be included in calculations. In any case, it's clear that cost comparisons cannot be made on the basis of average historical costs, precisely because the subsidies and social costs are so numerous, complex, and disputed. The only reasonable procedure is to do forward-looking calculations, which estimate the costs and benefits of nuclear energy versus other energy sources, assuming new plants are built today.

Such studies, carefully done, generally show that nuclear energy is markedly more expensive than the major alternatives at prevailing prices and reasonably anticipated interest rates.[9] The most recent, issued toward the end of 2004 by economists associated with the University of Chicago, found that the "levelized" cost of new nuclear power—that is, all capital and operating costs, discounted to present value at a range of interest rates—is between $47 and $71 per megawatt-hour, which is equivalent to 4.7 to 7.1 cents per kilowatt-hour. Electricity from coal-fired plants, by comparison, would cost $33 to $41; for natural gas, the range would be $35 to $45. Because of the high up-front capital costs of building nuclear reactors, overwhelmingly the most important variable factor influencing their costs is the long-term interest or discount rate. Accordingly, loan guarantees, accelerated depreciation, and investment tax credits are potent means of encouraging nuclear construction, and if the net social impact of building reactors is considered positive, a lower discount rate can be selected for analytic purposes to reflect environmental or health benefits.[10]

Estimates done at the Massachusetts Institute of Technology put the cost of nuclear electricity about 50 percent higher than electricity from fossil fuels. John Deutch, who was responsible for energy research under President Carter, and Ernest Moniz, who had essentially the same job under President Clinton, estimate the cost of nuclear electricity at 6.7 cents per kilowatt-hour, versus 4.2 cents for electricity from coal or natural gas, assuming current prices and no special accounting for environmental side effects. But, they conclude, "if a cost is assigned to carbon emissions…nuclear power could become an attractive economic operation. For example, a $50 per ton carbon value, about the cost of capturing and separating [but not storing!] the carbon dioxide product of coal and natural gas combustion, raises the cost of coal to 5.4 cents and natural gas to 4.8 cents."[11]

The University of Chicago's estimates assume that new nuclear plants would be based on more advanced technology, so that the first ones would cost somewhat more than those built later on. In the long run, they argue, "there is a good prospect that lower nuclear [levelized costs] can be achieved."[12] In other words, once utilities and manufacturers settled on some standardized designs and economies of scale were achieved, the cheapest nuclear plants might be strictly price-competitive with the more expensive power plants fueled by coal or natural gas. In an effort to expedite that kind of process, in the last decade, nuclear regulators in the United States and Europe have encouraged nuclear manufacturers to come up with standardized next-generation designs that would be easier to build than current plants and more inherently safe.[13] The emphasis is on passive features, which automatically, without human, mechanical, or electronic intervention, tend to stabilize a reactor if anything goes very wrong. Incrementally improved versions of both the boiling water and the pressurized water reactor have been approved on both sides of the Atlantic, so that, in principle, once a utility gets permission from local and national authorities to build a nuclear power plant at a certain location, construction can begin without further ado. Of course, since nuclear energy continues to be highly controversial, hardly anybody wants to go to the trouble of winning approval for new nuclear sites. If any new project is launched in the United States in the next few years—and a couple of groups are trying to do just that—the first of the next-generation reactors will most likely be built at sites already hosting nuclear facilities.[14]

Meanwhile, partly as a positive result of electricity deregulation and reorganization, some power producers in the United States have had considerable success standardizing the operation of existing reactors to improve both their reliability and their efficiency. Exelon and Entergy, notably, have bought reactors from utilities around the country and showed that an independent, profit-making company can make a good business out of managing reactors better than they had been handled previously.[15] To take one example, at Indian Point, just north of New York City on the Hudson River, Entergy has built a rather impressive new building just to house the staff running the reactor; beefed up the plant's simulation capabilities—it has a control room where all conceivable emergency contingencies are evaluated; and replaced many non-nuclear components of the reactors (turbine systems and the like), sometimes spending tens of millions of dollars for new equipment, to improve performance. It all goes to show that once a nuclear power plant has been built and is up and running, the facility can be economically managed to the owner's financial advantage.

The Indian Point reactor, because of its age and its proximity to New York City, has long been the subject of acute concerns about nuclear terrorism, in the sense of attacking a plant and detonating it in order to make it, in effect, a dirty bomb—a bomb designed to spread toxic radiation over a large area, but not actually to blow up like an atomic bomb. Naturally those concerns became all the more severe after 9/11, which raised the specter of a gang flying an airplane directly into a reactor containment building or of seizing adjacent territory and shooting shoulder-mounted missiles into the core.

Visit a plant like Indian Point,[16] or any similar nuclear power plant, however, and you will notice that a reactor is a hard target compared to many others a terrorist group could choose from. Even without all the special added security measures adopted since 9/11, the core is by nature highly protected by its pressure vessel and by the containment building, which is designed to prevent radiation from leading into the environment even in the case of a severe accident. The reactor vessel also is quite small—much harder to hit than a large, flimsy building like the World Trade Center towers, not to speak of sprawling petrochemical plants, trains and trucks carrying toxic or explosive materials, or transportation hubs. The worst plausible case seems to be that a terrorist group could sabotage water piping or controls entering the reactor vessel from the outside, inducing a meltdown. But the group would have to be exceedingly well informed about details of the reactor's design, and according to managers at Indian Point who have studied this kind of scenario, they could accomplish their ends only if a large commando group succeeded in seizing and holding territory near the plant, so as to mount a sustained military operation. It could happen, but the odds are remote, and much more remote than they are for many other plausible terrorist scenarios.

A more worrisome possibility is an attack on the ponds adjacent to most U.S. plants, where spent nuclear fuel is stored pending agreement on a method and site for final, permanent disposal.[17] That waste is highly radioactive—much too toxic, in fact, to be stolen and used as a source for the explosive material in a true atomic bomb—so if a spent fuel pond were bombed with conventional explosives, the result could be a dirty bomb of the worst kind. This consideration argues for transferring fuel from pools into dry-cask storage immediately after the spent fuel has cooled for the requisite five years, and for settling on how it can be stored for the long term or permanently.[18] In this procedure, fuel rods are kept in 6-meter-tall concrete-and-steel bins, either at the reactor site or in regional facilities. Two panels recently determined that dry-cask storage is quite safe and satisfactory for as long as 50 years, and some favoring a nuclear renaissance have argued that it would suffice to just keep storing fuel in dry casks at reactor sites indefinitely.[19]

In the early 1990s in the United States, it was decided more or less by administrative fiat that the best way to permanently store spent nuclear fuel would be at Yucca Mountain, a site in Nevada on the grounds where atomic bombs were tested in the 1950s and 1960s.[20] The subtextual thinking, so to speak, was that the site was already such a contaminated mess that the presence of well-contained spent fuel casks could hardly make things worse. Congress went along with the idea initially, promising the nuclear industry that a permanent disposal facility would be ready within a decade or so. Understandably, however, the citizens of Nevada vehemently disagreed with the whole idea and fought it tooth and nail, making their cause a tempting issue for national political candidates to embrace.[21] Matters were made worse by a fairly steady stream of reports finding that the site was not quite as geologically stable as first supposed—reports that the government, in some cases, tried to suppress or minimize.

Arguably, the whole subject of permanent nuclear waste disposal has been unnecessarily complicated by demands that a site be completely stable and uncontaminated for hundreds of thousands of years, which may simply be an unattainable standard. If wastes are stored retrievably, so that as technology evolves and more is learned about sites, they can be moved and re-stored, a central storage facility really does not have to be literally permanent. A 2005 report by the American Physical Society's public affairs panel concluded that "no foreseeable expansion of nuclear power in the United States" would require the country to drastically revise its plans for nuclear waste disposal. "Even though Yucca Mountain may be delayed considerably, interim storage of spent fuel in dry casks, either at current reactor sites, or at a few regional facilities, or at a single national facility, is safe and affordable for a period of at least 50 years."[22] But would it not be better, given the risk of terrorist attacks on temporary storage ponds, to agree on some one central site that can be very highly secured?

What certainly is not desirable as a supposed solution to the waste problem is reprocessing and recycling of nuclear fuels—an approach the nuclear industry has promoted in many countries, partly to create an impression that the disposal problem is solved, partly to stretch fuel resources. Extraction of reburnable uranium and plutonium from spent fuels, it is argued, greatly reduces the physical quantity of waste that must be permanently stored, and it would open the door to a so-called "plutonium economy" in which breeder reactors would run on the recycled fuel, producing more energy than they consume.[23] But recycling does not really solve the disposal problem: it merely reduces the volume of waste that has to be permanently stored (and that volume is relatively small to begin with), while in some ways complicating the whole situation by creating more streams of different radioactive materials that all have to be specially handled. As for the stretching of nuclear fuels, that benefit comes at the cost of having to widely transport fuels consisting of pure fissile material that could be ripe targets for terrorists seeking to build bombs.

Spent fuel from reactors, left alone, is too radioactive to be readily handled by a criminal gang, and extracting weapons-usable material from it would be beyond the capabilities of even an organization like Al Qaeda in the days before its large training camps were broken up. When plutonium is extracted from spent fuel commercially, however, it can be handled quite easily and could be used directly by a terrorist group or be stolen and sold to a government seeking to obtain nuclear weapons quickly and surreptitiously. This is why President Jimmy Carter was right to terminate all U.S. work on reprocessing and breeder reactors in 1977—a policy that should be rigorously upheld.

Some countries like Japan and France have continued to take the position that we need to reprocess spent fuels in order to stretch resources (and U.S. policy should continue to be to discourage them). Furthermore, some scholars have done calculations showing that if there were a huge expansion of nuclear energy, not only would uranium reserves be strained and the case for recycling become all the more tempting, but there also would be a tremendous need for fuel processing facilities, notably uranium enrichment plants. Since these are highly sensitive facilities and can be turned to the creation of explosive material for bombs—what Iran has been getting set to do[24]—their wide dissemination obviously is highly undesirable. But how much of a problem is this, really, and to what extent can it be prevented? Harold A. Feiveson of Princeton University, a specialist on nuclear energy and nuclear proliferation, has calculated that if the Intergovernmental Panel on Climate Change's projections of nuclear capacity are taken literally, the consequences are alarming. According to the IPCC's main business-as-usual scenario, nuclear energy will grow to 3,000 gigawatts by 2075—essentially the equivalent of 3,000 large nuclear power plants around the world. "The management of a nuclear system of 3,000 GW would be truly challenging," says Feiveson, with some understatement. "If based on a once-through fuel cycle using light water reactors, such a system would generate roughly 600 tons of plutonium annually, and would require on the order of one-half million tons of uranium annually. If based on liquid-metal plutonium breeder reactors, it would involve the fabrication into fresh fuel annually of over four thousand tons of plutonium."[25] The annual volume of highly radioactive spent fuel would be on the order of 50,000 to 70,000 tons, about the planned capacity of Yucca Mountain, and production of fresh fuel would require the construction of 200 large enrichment plants, on the same scale as the one operated by a German–Dutch–British consortium in the Netherlands.

This kind of calculation usefully establishes a sense of limits, but to consider the numbers is to realize that the business-as-usual scenario is not actually going to happen. Simply put, the world is not going to build 1,200 nuclear reactors in the next 50 years. The most rapidly developing economies, India and China, are building some reactors, but nuclear energy is basically not an attractive option for them because of the very high up-front costs. The European countries may build a fair number of replacement reactors in the next decades, but most of them hope to meet Kyoto requirements without having to add much nuclear capacity, mainly by building many large offshore wind farms.[26] So the question of whether a robust expansion of nuclear energy is necessary really is just an issue for two or three big countries, the United States foremost among them—the countries that may not be able to sharply reduce or even constrain carbon emissions otherwise.

There's a school of thought that any reliance on nuclear energy by anybody sets a bad example, and tempts wayward states to first acquire civil nuclear technology and then turn it to military ends. But this concern is hard to credit, on either theoretical or historical grounds. Is it really plausible that a country like North Korea or Iran would forego acquisition of nuclear technology just because the United States, Germany, or Japan decided to do so first? Obviously they have been assembling nuclear technology precisely in order to lay the groundwork for nuclear weapons programs: they're not buying it with peaceful purposes in mind, and only then contemplating the military option as an afterthought; they will remain intent on laying the groundwork for nuclear weapons programs, to the best of their ability, whatever other countries may do. While it's true that if every other country of the world gave up nuclear technology, it would be harder for countries like Iran and North Korea to launch weapons efforts in the guise of energy programs, it would be a very high price to pay for a very uncertain reward.

In the historical record, there is not a single clear case of a country buying civil nuclear equipment and only later opting for weapons, having been tempted into the decision by the availability of the technology. The United States, the Soviet Union, England, France, China, Israel,[27] Iraq,[28] India,[29] Pakistan,[30] and South Africa[31] *all* launched nuclear programs specifically to open a nuclear weapons option. (The only debatable case is India.) None of them would have been deterred by others claiming that reliance on nuclear energy is not desirable, per se.

Baldly put, boosting U.S. reliance on nuclear power so that it accounts for, say, 40 percent rather than 20 percent of our electricity would not materially affect the considerations of any other country deciding whether or not to develop nuclear weapons. (Nor, conversely, would a decision by the United States to end reliance on nuclear energy have any impact.) This does not imply, of course, that the United States should return to the reckless policies of the 1950s, 1960s, and 1970s,

when its government positively encouraged other countries to buy nuclear equipment.[32] On the contrary, the United States should continue
to work vigorously to convince the other major countries that supply
nuclear equipment that they should limit sales of sensitive technology
to dubious customers like Iran and North Korea.

The zeal and stubbornness with which opponents of nuclear energy
have sought to guarantee that it can't work is sometimes puzzling, even
taking all the horrors of nuclear weaponry into account. When people
are called upon to weigh nuclear risks against other risks at the grass-
roots level, otherwise reasonable people seem to shy away from the effort. The experiences of John Weingart, an idealistic specialist in public
health and environmental policy who has had a succession of increasingly responsible positions in the state of New Jersey, are typical. As a
student at Princeton University's Woodrow Wilson School in the early
1970s, Weingart was known for his wry sense of humor as well as his
sense of public service—not to mention a Sunday evening radio show
he did, and continues to host, devoted to unusual American folk music
("music you won't hear on the radio"). In mid-career, Weingart found
himself saddled with the thankless job of trying to find a home for low-
level wastes from the state's nuclear power plants—wastes that are not
terribly radioactive or toxic, and not all that different from wastes routinely generated in medical facilities around the world. The thinking
was that the problem could be solved by providing some community
with suitable economic incentives, but Weingart discovered after four
unpleasant years of hard work that in fact no community was willing to put itself through the process of public debate that would have
been required to reach a decision. Having described his experiences in
a sardonic report, "Waste Is a Terrible Thing to Mind,"[33] Weingart says
the whole business left him with views about nuclear energy that are
probably closer to President Bush's than to those of "the liberal groups
I used to give money to."[34]

Thus we confront "a paradoxical situation," observes Richard Meserve,
a nuclear physicist and attorney who served as chairman of the Nuclear
Regulatory Commission under President Clinton. "Those who should
be the strongest advocates of nuclear power—environmentalists, governmental policy makers concerned about global warming, and generating companies with an economic stake in nuclear's future—are unable
or unwilling to advance the most compelling argument in support of it.

Without advocacy by those who see the benefits of nuclear power, it is only to be expected that full exploitation of the nuclear option will be limited or deferred indefinitely."[35]

Despite Meserve's well-founded concerns, what has been a politically frozen situation does show some sign of thawing. From James Lovelock, the father of the Gaia hypothesis that treats all of earth's systems as a living whole, to environmental visionary Stewart Brand, calls are being issued to reconsider the nuclear option. The same message has been coming out of some big collaborative policy exercises, from the National Energy Policy Initiative associated with Amory Lovins's Rocky Mountain Institute to the National Commission on Energy Policy, an initiative launched by leading liberal-minded foundations in 2002.[36] When *The New York Times*'s well-informed and open-minded columnist Nicholas D. Kristof wrote an opinion piece in early 2005 saying that "nuclear energy seems much safer than our dependency on coal, which kills more than 60 people a day," the paper received a torrent of letters supporting his position.[37]

In Sweden, a majority of the public is reported to be having second thoughts about the country's pledge to renounce nuclear energy. Not only is a phaseout easier said than done but also, Swedes may be aware that in countries where similar commitments have been made, ironic consequences are sometimes evident. Austria, having said no to nuclear, now finds itself importing nuclear-generated electricity from Czechoslovakia, its immediate neighbor to the northeast. The Czech electricity is produced in reactors that may be inferior to the latest models; in any case, the reactors certainly are not subject to whatever safety standards Austrians would want to enforce. In the United States, some polls suggest that as much as two thirds of the general public is ready to embrace nuclear energy.

To be sure, among those opposing the nuclear option and preferring alternative approaches to carbon abatement, there are still those building straw men: in another opinion piece that appeared in the *Times*, a peace activist and a research manager joined forces to deliver a resounding rejection of nuclear, on the grounds that meeting anticipated growth in U.S. energy demand over the next 50 years would require construction of 1,200 new nuclear power plants, in addition to the 104 already operating.[38] But there need not be and should not be any growth in U.S. energy demand in the next 50 years. Rather, what needs to be done immediately is to eliminate gratuitous greenhouse gas emissions from the readily identified dirtiest sources. A combination of gasoline taxes, stricter fuel-efficiency standards, and carbon limits, set at the maximum level the public will accept, can get part of the job done. But the other part is best done by shuttering the oldest, least efficient, and dirtiest coal plants not operating, and replacing them with some combination of natural gas and nuclear plants.

With tech-savvy but deliberately provocative magazines like *Wired* publishing articles with titles like "Nuclear Now: How Clean, Green Atomic Energy Can Stop Global Warming,"[39] this may sound like jumping on the bandwagon. But consider the economics at the simplest level. In the compromises on air quality regulation promulgated by the Bush administration in early 2005, the cost of cleaning up the nation's dirtiest coal-fired plants was estimated to be in the vicinity of $7 billion. Does it make sense to spend all that money to scrub pollutants from a group of very old plants that will continue, even after the cleanup, to emit carbon dioxide in staggering quantities? It would be better to just replace those plants with nuclear reactors, which could be deployed at a total cost of perhaps $50 to $100 billion—not a big bill for a country like the United States—and would generate pollution-free, carbon-free electricity for at least 60 years thereafter.

Conclusion

How to Reduce Greenhouse Gases Now, Using Today's Technology

PREDICTIONS ARE by nature treacherous. The Greek historian Herodotus, in his account of the Persian wars, tells a famous cautionary tale. There lived in the sixth century B.C. a monarch named Croesus, who ruled the prosperous kingdom of Lydia, in what is now western Turkey. The Lydians, Herodotus tells us, were in almost all important respects culturally indistinguishable from the Greeks on the other side of the Aegean. Croesus, having ascended to the throne in 560, having proceeded to subjugate many neighboring peoples, and having amassed the wealth with which his name has been synonymous ever since, decided to challenge Cyrus, king of the Persian state emerging to his northeast. Being a good Greek, Croesus consulted the oracle in Delphi, and his messengers brought back the following prophecy: if he made war on the Persians, a great empire would be destroyed. Croesus went to war and was decisively defeated by Cyrus at Sardis in 546, and a great empire was destroyed—his own. The oracle's prediction had been correct; only the inference drawn by Croesus was wrong.[1]

When scientists tell us today that the earth is destined to become a few degrees warmer in this century, and very likely more than that if concerted action is not taken right away, it doesn't at first sound like such a big deal. Certainly for anybody living in remote Siberia or on Tierra del Fuego, a few extra degrees of warmth must seem a definite improvement. Even in more temperate climes, like New England or Scandinavia, where spring and fall are painfully short, a great many people must think of slightly higher global temperatures as portending a better-balanced four-season year.

One rather obvious point that's overlooked in that sense of things is the averageness, if you will, of what's being predicted. When scientists say the world will be 3° C warmer 75 or 100 years from now—or 5 or 6 degrees, Fahrenheit—of course they do not mean it will be 3° C warmer everywhere and at all times. Some places will be much hotter than that some of the time, and other places, paradoxically, may actually be colder some of the time. Even if the average increase in some particular region is modest or below average, the range between the hottest and coldest days of the year may become wider, and extreme climate events may become more frequent. The world got a glimpse of what's involved in the summer of 2003, when an unprecedented heat wave engulfed Western Europe, sending temperatures well over 100° F in countries where people ordinarily have no need to own air conditioners and have little inkling of how dangerous excessive heat can be to the old and infirm. In France, where the elderly live unusually long and often alone, about 12,000 people are thought to have perished from exposure.

Among the recent events that have taken even climatologists by surprise has been the extent of ice thinning in the area around the North Pole, which has warmed much faster than the world as a whole. One important reason the Arctic warming has been faster than generally expected is an insidious feedback mechanism: as ice melts and turns to seawater, instead of reflecting sunlight, it absorbs it, so that the pace of warming accelerates. Another insidious feedback mechanism, also experienced largely in the northern regions of the Northern Hemisphere, is that as the land surface gets drier with warmer temperatures, wildfires catch and spread, sending still more carbon dioxide into the atmosphere. In the summer of 2003, Siberian wildfires reportedly incinerated areas almost the size of Oregon.

In hindsight, such feedback mechanisms often seem obvious, but beforehand they are not always easy to anticipate, partly because there are so many possible linkages. The gravity of the effects also can be hard to appreciate ahead of time. A warmer Alaska sounds like a good idea in principle, but in reality it's turned out to be a mixed blessing at best. Besides having a devastating impact on some habitats, melting permafrost has been causing oil pipelines to sink and split, and the foundations of whole communities to come apart. So, even if the general trend of greenhouse gases and temperatures is gradually upward, and the consequences of climate change are also largely linear, those effects can add up and combine in ways that are not easy to foresee or to prepare for adequately.

In the fall of 2005, as this book went to press, news continued to accumulate that global warming, even if incremental, still could be a very serious matter indeed. Climatologists at NASA's Goddard Institute for Space Studies in New York—Jim Hansen's outfit—reported that 2005 was on track to be the warmest year yet recorded. Shrinkage of the Arctic ice cover was reaching unparalleled proportions, and the seasonal surface melt in Greenland—which has been observed from satellites since 1979—was seen to be accelerating. There were record-high ocean temperatures in the Gulf of Mexico, which energized the hurricanes that devastated New Orleans and threatened Texas.

Researchers at Purdue University, basing their assessment on what they said was the most comprehensive climate model yet employed for the United States, reported in October 2005 that the country can expect more extreme temperatures just about everywhere as this century wears on, as well as more extreme precipitation along the Gulf Coast, in the Pacific Northwest, and east of the Mississippi. Taking more factors into account than previous models have—the reflective effects of snow, for example, and the weather-blocking effects of mountains—the Purdue team said that the Southwest, where summer temperatures already exceed 115° F, will experience heat waves of even greater intensity and that the gulf will have larger volumes of rainfall concentrated in shorter periods of time. "The changes our model predicts are large enough to substantially disrupt our economy and infrastructure," said Noah S. Diffenbaugh, the Purdue team's leader.[2]

On December 1, 2005, scientists at Britain's National Oceanography Center reported in *Nature* that the Atlantic currents considered partly responsible for warming Northern Europe have weakened by a third. Some models, including those developed by Manabe and his protégés in Princeton, have indicated that faltering of the currents could induce abrupt European cooling. Meanwhile, scientists at the Max Planck Institute for Meteorology in Hamburg, Germany, predicted that the climate will change in this century more quickly than it has at any time in the recent history of the earth. They said sea levels could rise as much as 30 centimeters, with temperatures going up 4° C. European summers will be drier and winters warmer and wetter.[3]

Assessing the expected results of global near-surface warming trends, researchers from the Scripps Institution of Oceanography and the University of Washington found that in a somewhat warmer world, less winter precipitation would fall as snow, and the winter melt would occur earlier in spring rather than in summer and fall, when demand for water is highest. "With more than one sixth of the earth's population relying on glaciers and seasonal snow packs for their water supply, the consequences of these hydrological changes for future water availability…are likely to be severe," they concluded, largely confirming the findings of Manabe more than 20 years before.[4] In another article in the same issue of *Nature*, researchers with the U.S. Geological Survey anticipated rather drastic effects on river flows and water runoff, also consistent with the models developed by Manabe and his colleagues. Basing their analysis on an ensemble of 12 computer runs, they found that relatively wet regions will get wetter and dry regions drier: they predict runoff increases ranging from 10 to 40 percent in eastern equatorial Africa, the La Plata region of South America, and in the northern reaches of North America and Eurasia; and decreases of runoff ranging from 10 to 30 percent in southern Africa, southern Europe, the Middle East, and the middle regions of the North American West.[5]

Those kinds of effects, all more or less linear extrapolations from changes already being observed all over the world, are grave enough. But what if, as carbon dioxide levels double, triple, or quadruple, the effects are not merely linear or "monotonic," as some scientists like to say? What if, as greenhouse gas concentrations increase, there is some kind of unforeseeable change in the climate system state—something as severe as the shutdown of the North Atlantic conveyor postulated by Broecker and others, but possibly entirely different in detail? As temperatures go up in lockstep with greenhouse gas concentrations, reaching levels unprecedented not only in recorded history but also in the paleohistory extracted from ice cores and ocean sediments, can the possibility of radical changes in ocean-atmosphere dynamics be excluded?

Very likely Manabe is correct when he argues that the consequences of higher temperatures will not be as drastic as those from lower temperatures—that is, just because a 50 percent drop in greenhouse gas levels was enough to somehow induce ice ages in the past, this does not mean that a 100 percent increase will do something just as extreme in the future (see chapter 6). But Manabe may not be correct. Perhaps a doubling, tripling, or quadrupling of greenhouse gases will have consequences that are flatly catastrophic, and on a supraregional scale. The material foundations of whole civilizations could be threatened.

The United States is by no means immune to the dangers of regional climate disasters, as Hurricane Katrina recently showed. Some of the country's leading scientists are among those who have proved beyond a reasonable doubt that cataclysmic climate changes are known to have occurred abruptly in circumstances much like those we live in today (see chapter 7). Yet the U.S. citizenry does not seem very aware of such possibilities, and the notion that the nation should do something substantial to reduce the probability of climate catastrophe is widely seen as some kind of moral imperative, as if it were something Americans were being asked to do for somebody else.

The specter of a climate cataclysm seems to be more deeply appreciated in Europe—not to mention in the island states of the Pacific and in low-lying countries like Bangladesh—and doing something to reduce the probability of disaster is accepted as a simple, practical matter. In Holland, where control of water always has been fundamental to the nation's prosperity and even survival, water planners know well that increased Alpine runoff into the Rhine River system could leave the country submerged. Everybody in the Netherlands grasps that rising ocean levels caused by global warming could have the same effect. Huge barriers have been built to keep the oceans at bay, not only along the Dutch coasts but also across the English Channel in Great Britain's Thames estuary.[6] Throughout Northern Europe, every grade schooler learns of the region's dependence on the Gulf Stream for warmth and moisture—and everybody knows how serious any disruption to the oceanic conveyor could be.

Accordingly, the debate over global warming and the requirements of the Kyoto Protocol has had a very different tone in Europe than in the United States. Although thoughtful Europeans concede that Kyoto is but a first step and that its limits are somewhat arbitrary, the European arguments have been by and large not over whether to stick with Kyoto, but rather about how to meet its requirements and what to do next. In the United Kingdom, as this book went to press, the Labor government was considering—at the behest of the prime minister's science adviser—whether a renewed program of nuclear construction would be necessary to meet Kyoto commitments. Germany, still committed in principle to meeting Kyoto requirements with aggressive construction of wind turbines, elected a new conservative government inclined to reconsider its planned nuclear phaseout. (Chancellor Angela Merkel cautiously suggested in the campaign that the country should seek "an exit from the nuclear exit.") Yet Britain and Germany were much closer to meeting their Kyoto targets than some of the other European nations; Spain, Greece, Portugal, Ireland, and Finland were all way over targets—and informed people in those countries were seriously worrying about how to change course.

At the beginning of 2005, the European Union introduced a carbon trading system, covering some 12,000 installations designated major sources of carbon emissions. The system enables owners of those installations to purchase and trade emissions allowances, so as to meet the union's overall ceiling in the Kyoto regime. Within months of its inauguration, the price of a carbon allowance jumped from less than 7 euros per metric ton of carbon dioxide to about 20 euros/ton.[7] Yet sympathetic critics of the system still wondered whether the price would go and stay high enough to prompt long-term investments in carbon mitigation technologies. Some considered the system unfair because only industrial emitters are subjected to carbon caps, while the automotive, home, and agricultural sectors are left unregulated.

Partly because of such considerations, Japan's environmental ministry was taking the position at the end of 2005 that meeting Kyoto commitments would require imposition of a carbon tax by 2007. A carbon tax has the virtue of affecting all players equally, penalizing them exactly to the extent they contribute to global warming, but this approach has not found much favor in the United States. (The conservative jurist Richard Posner is an exception, as noted in chapter 8, in advocating such a tax as the most effective means of reducing the likelihood of a climate catastrophe.) The cap-and-trade system pioneered in the U.S. acid rain program is much preferred: California and the northeastern states have launched regional programs to introduce such systems for carbon, and also are taking measures to mandate improved automotive fuel efficiency and reduced automotive emissions. At the same time, U.S.–based multinational companies are bound to be drawn into carbon trading systems, if only because some of their subsidiaries are located in regions requiring participation in such systems.

At the beginning of 2005, the Kyoto Protocol formally took effect with the accession of Russia, and it provides powerful incentives to participate in a global carbon trading system going well beyond the intra-European and other regional systems. Like the sulfur trading system pioneered in the United States, the Kyoto system promises to greatly reduce the overall costs of cutting greenhouse gas emissions. But without the United States as an active participant, buying emissions permits from countries more easily able to meet Kyoto requirements, the world carbon price will not be as high as it otherwise would be. An important incentive to everybody to reduce emissions will be weakened, because of U.S. intransigence.

Some U.S. states, municipalities, and even companies are taking measures to comply with the Kyoto regime (see "You *Can* Do It at Home"). But the policy of the current administration in the United States has been to stay aloof from Kyoto, and instead to emphasize long-term development of new energy technologies and new means of capturing and storing carbon. To put that approach in the most positive light, there is indeed every reason to promote development of photovoltaics, fuel cells, and equipment using the high-temperature superconductors. Those technologies all have the potential to revolutionize energy in the next generation. So, too, do clean coal technologies and carbon sequestration.

But none of these technologies is close to commercial fruition today, so to base carbon mitigation strategy on them is to put the cart before the horse. Partly because of the term "precommercial," often used to describe such technologies, many people mistakenly believe that the technologies inevitably will be economically viable in some foreseeable period of time. But in fact, a precommercial technology is by definition a technology that has yet to be commercially proven—and that may *never* be proven. One might reasonably suppose, for example, that since the cost of making photovoltaic cells has been steadily coming down, eventually electricity from such cells will be competitive with electricity from coal, gas, or nuclear reactors. Right now, as shown in chapter 9, photovoltaic electricity is ten times as expensive as electricity from wind, which is just becoming commercially viable. But won't photovoltaic electricity soon become just as cheap as wind, or even cheaper? The answer, regrettably, is not necessarily. In the first place, it isn't a foregone conclusion that photovoltaic costs will just keep coming down and that efficiencies will keep rising. And second, even if they do, the costs of all competing electricity technologies also will continue to fall, so that the future competitiveness of solar is not guaranteed.

When the talk is of the bundle of technologies required to make the hydrogen economy work (see chapter 10) or all the technologies involved in the capture and storage of carbon from emissions (see chapter 8), the future is even more uncertain. Many technologies must come to commercial fruition more or less simultaneously for the hydrogen economy or carbon sequestration to work, and if just one fails, the whole interdependent enterprise falters.

This book has argued that global warming represents a kind of international emergency, requiring immediate concerted action. As greenhouse gases approach levels never seen in the last 700,000 years, we are drifting into uncharted waters, and so the rational thing to do is reverse course and get back out, the faster the better. The United States, responsible for a quarter of the world's greenhouse gas emissions, the world's richest and most capable country, and the country that uses energy the most extravagantly, should take especially aggressive action to reduce emissions. Rather than being a global laggard, the United States should be a world leader in this effort.

In principle, as other countries already have shown, emissions can be reduced either by means of a cap-and-trade system, with special provisions for the automotive sector, or by means of a carbon tax. In terms of ultimate result, it doesn't much matter which way the United States goes. But both elementary equities and administrative simplicity favor a carbon tax. Everybody is affected equally by such a tax, to the exact extent they are responsible for carbon emissions. The government does not have to get into the business of picking technological winners or losers (which is why conservatives like Posner like this approach), and it can confine and concentrate its activity, once the tax is in place, to mitigating the regional and socioeconomic effects of the tax wherever they prove most adverse (which is why liberals and social democrats should like this approach too).

In the automotive sector, to the extent that achieving a greater degree of independence from oil exporters is considered an imperative, a higher gasoline tax could be imposed on top of the carbon tax. But there are limits to how far such policies can be carried, as a matter of political realism and simple equity. The United States is profoundly dependent on automotive transport and, because of its huge extent and sprawling suburban style, always will be. It will not be possible to persuade the public to accept gasoline taxes as high as those imposed in the European countries and Japan, and such taxes should not be imposed. They would be unfair to lower-income workers who depend on their cars to get to work and to transact family business, and there really is no way around this dilemma: if money spent on gasoline were somehow rebated to society's less advantaged members, much or most of it would just be used to buy more gasoline. (Just as people drive more miles and build bigger houses when energy prices are lower, so that they end up using just as much energy as before, given a rebate for energy use, they'd use it to buy more energy.)

This is why, in a country like the United States, cuts in greenhouse gas emissions must be made across the board. In the electricity sector, a high enough carbon tax will encourage conservation, so that overall usage stays level even as the economy grows, and will prompt rapid substitution of low-carbon for high-carbon fuels. Instead of continuing to focus an enormous amount of public attention and political capital on the vexing issue of how to clean up the nation's dirtiest coal-fired power plants, for the sake of public health, they will simply be shut down. Depending on how the economics of individual technologies and public preferences evolve, they will be replaced by some mixture of wind turbines, natural gas, and nuclear reactors.

None of those technologies will require any government subsidies or special breaks; termination of the Price–Anderson program, which limits corporate liability in nuclear accidents, is long overdue. If nuclear reactors now work as well as the industry and its regulators claim, the industry should be able to insure itself. Once coal-fired generation is properly priced to reflect its huge social costs, wind also will be able to compete without the public favoritism it still depends on everywhere.

Making use of the carbon abatement wedges described in chapter 9—seven of which are needed to keep greenhouse gas levels constant for the remainder of this century—the probable effects of a U.S. carbon tax can be estimated roughly as follows (see graph). Conservation of energy in the industrial, commercial, and residential sectors, as well as in the power sector itself (by means of combined heat and power), ought to provide two wedges. Reduced use of gasoline in the automotive sector could account for one and a half wedges. Aggressive deployment of wind turbines and solar technologies can provide at least one and a half wedges, and substitution of nuclear energy and natural gas for coal in the power sector yields the final two wedges.

When the global picture of carbon abatement wedges is translated into concrete, U.S. terms, it looks like this: coal's role in the generation of electricity is cut from about 55 percent roughly in half, so that it now constitutes around 25 percent. Nuclear's contribution more or less doubles from 20 to 35 or 40 percent, and some combination of wind and natural gas makes up the remaining 15 percentage points. Automotive use of gasoline is cut perhaps by half, as a result of improved technology and higher taxes (admittedly an optimistic projection). Across-the-board conservation prevents energy usage as a whole from growing. The bottom line is that U.S. greenhouse gas emissions, in 10 or 15 years, are cut by a quarter to a third, bringing the United States into step with the Kyoto program.

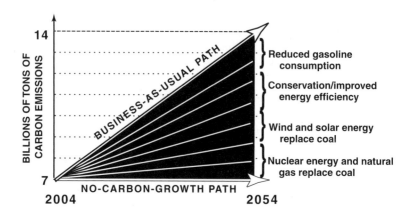

Breaking the Carbon Habit

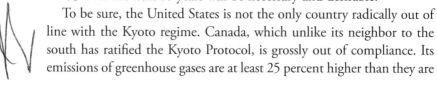

Aggressive expansion of wind energy means that many hilltops, ocean vistas, and open prairies will soon be dotted with large steel turbine towers—much of the United States will have the same kinds of landscapes seen today in northern Germany and Denmark. To the extent that natural gas is substituted for coal, at least one new major pipeline will have to be built from Alaska to the Lower 48, and members of coastal communities will have to accept construction of terminals to receive liquefied natural gas. (Seen in a global perspective, Pacala and Socolow estimate that achieving one of their carbon-mitigation wedges by means of substituting natural gas for coal would require building one Alaska pipeline per year for 50 years or docking and discharging 50 large CNG tankers each day.)

Even allowing for that rapid construction of wind turbines and some added reliance on natural gas, it will not be possible to replace the country's dirtiest and most carbon-intense coal-fired generators without substantially increasing the size of the U.S. nuclear reactor fleet. An increase in the total U.S. complement of reactors from 100 today to 150 or 175 over the next 10 years will be necessary and desirable.

To be sure, the United States is not the only country radically out of line with the Kyoto regime. Canada, which unlike its neighbor to the south has ratified the Kyoto Protocol, is grossly out of compliance. Its emissions of greenhouse gases are at least 25 percent higher than they are

You *Can* Do It at Home

Not content to wait for the U.S. government to join international efforts to curtail greenhouse gas emissions, 9 northeastern states adopted a draft plan in August 2005 to cap and then reduce carbon emissions from 600 power plants. Acting on the initiative of George E. Pataki, the Republican Governor of New York State, the states would impose a ceiling on total emissions of 150 million tons, starting in 2009. That would be enforced until 2015, whereupon Connecticut, Delaware, Maine, Massachusetts, New Hampshire, New Jersey, New York, Rhode Island, and Vermont would systematically reduce emissions by 10 percent over the following 5 years. Together, these states account for about the same quantity of carbon dioxide emissions as Germany.

Separately, Republican Governor Arnold Schwarzenegger has ordered that California cut power plant emissions back to 1990 levels by 2020, with a further cut of 20 percent by 2050. Cars sold in California will have to meet much tougher tailpipe emissions standards by 2009, and at least 10 percent of the cars sold in that model year will have to be zero-emissions vehicles—automobiles powered by batteries or fuel cells.

As the Kyoto Protocol entered force in February 2005, and the members of the European Union launched a cap-and-trade system affecting 12,000 facilities, European businesses grumbled that they were going to be at a competitive disadvantage versus corporations based in countries that were not party to the protocol. But the leaders of some of those supposedly advantaged multinational and U.S. corporations worried they might be left behind in the long run, as others get a head start in the business of trading carbon credits. Accordingly, a group of companies that included DuPont, International Paper, Ford Motor Company, American Electric Power, and Motorola established a pilot Chicago Carbon Exchange in 2003. Trading has grown briskly on that exchange, though the value of credits has remained lower than in Europe, as their usability may be limited in the foreseeable future to the northeastern states and California.

Among the pioneering buyers of emission credits has been TransAlta Corporation, a Canadian utility that ranks as that country's second-largest source of carbon dioxide emissions. TransAlta contracted to buy 1.75 million global warming credits over 10 years, in effect enabling it to reduce its own emissions in Canada, Mexico, and Australia considerably less than otherwise would be required, by investing in a project to reduce methane emissions from pig manure at a very large Chilean hog farm. It is this kind of deal that could lead to substantial cuts in emissions in countries like China and India, if all the nations of the world were fully participating in a global trading system.

On December 12, 2005, *BusinessWeek* magazine published a study of what individual corporations in the United States and around the world were doing on their own initiative to reduce their greenhouse gas emissions. Topping the list was the U.S. company DuPont, which cut its energy use 7 percent below 1990 levels and its emissions by 72 percent. "Back in the mid-1980s," the magazine commented, "DuPont created a profitable business selling substitutes for chlorofluorocarbon (CFC) refrigerants that were destroying the earth's protective ozone layer. Tackling climate change was a natural extension of that experience." Other top performers—according to a survey that admittedly depended largely on self-reported, unaudited, and sometimes anecdotal information—included British Petroleum, Bayer, British Telecom, and Alcoa.

Of course, what's motivating most of these companies is not merely high-mindedness or public relations but, rather, a firm expectation that much sharper carbon limits are merely a matter of time. In June 2005, reversing an earlier stance, the U.S. Senate voted 54–43 for a resolution calling for enactment of a "comprehensive and effective national program of mandatory market-based limits and incentives on emissions of greenhouse gases." Companies that would be most affected by such limits, according to the *BusinessWeek* survey, included both the top U.S. coal-burning utilities (AEP, Southern, Duke, and Cinergy), which have largely accepted the inevitability of carbon caps, and the top U.S. oil companies (Exxon Mobil and ChevronTexaco), which have not. The oil companies typically emit about as much greenhouse gas in their exploration and extraction operations as the utilities do making electricity.

supposed to be, just as if it had refused, like the United States, to ratify in the first place. Australia, like the United States a lone dissenter from Kyoto, decided in 2005 to terminate subsidies that had encouraged rapid deployment of wind turbines. Like the United States, Australia was moving backward, bequeathing to itself a much tougher task if and when it decides to join the rest of the world in curtailing emissions.

Many major countries of the world, however, are making a serious and sincere effort to meet Kyoto requirements. From their point of view, the United States is getting a free ride—benefiting from the costly and painful cuts in emissions that they are undertaking and, by not fully participating in the Kyoto cap-and-trade system, crippling its full effectiveness. For those countries, perhaps the most difficult decisions ahead will concern what to do about the ones rejecting the agreement.

Right now, any talk of penalties is still unthinkable. But as evidence of global warming and adverse effects continues to accumulate at an acclerating rate, that too may change.

ACKNOWLEDGMENTS

Too many people were interviewed for this book to be thanked individually. I do, however, want to acknowledge the following scientists and scholars who were exceptionally generous with their time and attention: Wallace S. Broecker, Curtis Covey, Paul Crutzen, Willi Dansgaard, James Hansen, John A. Harper, Philip H. Heckel, Sigfus Johnsen, Chester C. Langway Jr., Lester Lave, Jerry D. Mahlman, Syukuro Manabe, Veerabhadran Ramanathan, and Bernhard Stauffer.

The Spieler Agency's John Thornton provided untiring support and valuable editorial guidance at every stage of the book's preparation, as did the team at Columbia University Press—Jim Jordan, Patrick Fitzgerald, Leslie Kriesel, and Ann Young. I'm grateful to Susan Hassler, the editor-in-chief of *IEEE Spectrum* magazine, for helping me with some aspects of the book proposal and giving me time off to bring the project to completion. Prachi Patel, an editorial intern at *Spectrum* in 2004–05, read the entire manuscript and made many helpful suggestions. My daughter, Anna Robinson-Sweet, helped with illustrations, and my friend Norman Filzman with proofing.

Most of all I want to thank my brother-in-law, Walter Robinson, an atmospheric scientist at the University of Illinois, Urbana–Champaign, without whom this book could not have been written, let alone conceived.

More than the usual disclaimer is in order: not only are any faults in the book wholly my responsibility, but none of the individuals listed above would agree with every aspect of the argument.

The book is dedicated to my wife, Gail Robinson, with the same disclaimer about the final contents. She provided expert editorial guidance as well as substantive suggestions—and disagreed constructively with much of what I have to say.

NOTES

PREFACE

1. Excellent general treatments of global warming and climate policy include Sir John Hougton's *Global Warming: The Complete Briefing* (Cambridge University Press, 3rd ed., 2004) and William Ruddiman's textbook, *Earth's Climate: Past and Future* (W. H. Freeman, 2001). For somewhat more technical treatments, the Intergovernmental Panel on Climate Change's periodic reports are well worth consulting. The next major IPCC report will be released early in 2007.

1. THE CASE FOR SHARPLY CUTTING
U.S. GREENHOUSE GAS EMISSIONS

1. In research published as this book was going to press, Kerry Emanuel of the Masachusetts Institute of Technology determined that average hurricane intensity has increased 50 to 80 percent in the last 30 years. Peter Webster of Georgia Tech and Greg Holland at the National Center for Atmospheric Research found that the frequency of category 4 and 5 storms has nearly doubled.

2. United Nations Development Program, *2005 Human Development Report*, Table 22, p. 289.

3. World Bank, *2005 World Development Indicators*, Table 3.7.

4. United Nations Development Program, *2005 Human Development Report*, Table 22, p. 289; for comparison, see World Bank, *2005 World Development Indicators*, Table 3.8.

PART 1. COAL: A FAUSTIAN BARGAIN
WITH PAYMENTS COMING DUE

2. THE BASIS OF IT ALL:
PENNSYLVANIA IN THE PENNSYLVANIAN

1. The Philadelphia newspaper is quoted in Freese, *Coal*, 107.

2. Ibid. 106. It was Lyell who enunciated the fundamental "uniformitarian" principle that the same underlying physical processes account for all the world's geological phenomena. That principle, considered the bedrock of modern geology, implied that earth's past, present, and future climates, as well as its crust, could be understood in essentially the same terms.

3. As late as the American Revolution, not long before Lyell's visit to the United States, the director of France's natural history museums in Paris, Comte Georges de Buffon, was driven from the city because he dared suggest that earth might be as old as 70,000 or even 100,000 years. Buffon's exile is one of the first facts to be highlighted for visitors to Paris's Museum of Mineralogy. One of the most powerful geological tools of all, the method of correlating rock strata by reference to the fossils found in them, was devised by a contemporary of Buffon's, Georges Cuvier, who had discovered evidence of species extinctions in the sedimentary rocks of the Paris basin. That technique was the first and is still the most important method used to tease the history of the world's climate out of what's seen on or near the earth's surface. See Hallam, *Great Geological Controversies*, chapter 2, and especially page 38.

4. Labor admittedly did not obtain full recognition in the process that led to the settlement, but Roosevelt basically took its side, and this "square deal" was the first big victory for the American Federation of Labor. See Joseph G. Rayback, *A History of American Labor* (New York: The Free Press, 1966), 212, and George E. Mowry, *The Era of Theodore Roosevelt and the Birth of Modern America* (New York: Harper & Row, 1958), 134–41.

5. The author toured Pittsburgh road cuts with John Harper on October 18, 2003.

6. For a schematic of rock successions of the type described here, see figure 1 in Richard M. Busch and Harold B. Rollins, "Correlation of Carboniferous Strata Using a Hierarchy of Transgressive-Regressive Units," *Geology* 12 (August 1984): 471.

7. McPhee, *Annals of the Former World*, 93–94. Similar descriptions are to be found in Winchester, *The Map That Changed the World*, 47–49 and 173–75, and Freese, *Coal*, 17. The phrase "Pennsylvania in the Pennsylvanian," used in the title of this chapter, is McPhee's.

8. The midcontinent sea is described in Philip H. Heckel, "Glacial-Eustatic Base-Level-Climatic Model for Late Middle to Late Pennsylvanian Coal-Bed Formation in the Appalachian Basin," *Journal of Sedimentary Research* B65 (3) (August 15, 1995): 351; and "Evaluation of Evidence of Glacio-Eustatic Control over Marine Pennsylvanian Cyclothems in North America," *SEPM Concepts in Sedimentology and Paleontology* (Society for Sedimentary Geology) 4 (1994): 79.

9. See for example James Lee Wilson, "Cyclic and Reciprocal Sedimentation in Virgilian Strata of Southern New Mexico," *Geological Society of America Bulletin* 78 (July 1967): 805–17; also D. R. Boardman and Philip H. Heckel, "Glacial-Eustatic Sea-Level Curve for Early Late Pennsylvanian Sequence in North-Central Texas," *Geology* 17 (1989): 802–5.

10. The Exxon stratigraphy was associated with names like Peter Vail and John Van Wagoner.

11. Heckel, "Evaluation of Evidence of Glacio-Eustatic Control," 65, 81.

12. The seminal article: Harold R. Wanless and Francis P. Shepard, "Sea Level and Climatic Changes Related to Late Paleozoic Cycles," *Bulletin of the Geological Society of America* 47 (August 31, 1936): 1177–1206. Background to the story of how their hypothesis won the day was obtained in interviews with Heckel, October 8, 2003, and Rollins, October 31, 2003.

13. Du Toit was closely associated with the controversial theories of continental drift, advanced by the German Alfred Wegener, who postulated around the time of World War I that during the late Paleozoic all the continents had been joined in one supercontinent, Pangea. While Wanless and Shepard noted that typical glacial deposits such as tillites and varve slates also had been found in South America, Australia, and even India, they were careful to dissociate their glacial-eustatic model from Wegener's, which they considered inessential. And they made no mention of Gondwanaland, the southern supercontinent postulated by Wegener's Austrian contemporary Edward Suess, who based his notion on certain fossils and animals found in the coal-bearing deposits of India's Gondwanaland District by British-trained Indian geologists. See Hallam, *Great Geological Controversies*, 112.

14. Wanless and Shepard, "Sea Level and Climatic Changes Related to Late Paleozoic Cycles," 1194.

15. J. C. Crowell, "Gondwana Glaciation, Cyclothems, Continental Positioning, and Climate Change," *American Journal of Science* 278 (1978): 1345–72; J. J. Veevers and C. McA. Powell, "Late Paleozoic Glacial Episodes in Gondwanaland," *Geological Society of America Bulletin* 98 (April 1987): 475–87.

16. Crucial to the acceptance of plate tectonics were the highly precise radio spectrometry techniques invented around World War II, which permitted the exact characterization of a rock's magnetic properties. By means of the technique, rock strata could be correlated in space and time, and it became irrefutable that continents now separate once were linked.

17. The discovery of so many depositional sequences, all seemingly depending on similar and concurrent processes of subsidence and deposition, suggested "that the causative processes were systematically linked," and over very wide areas, as Wilson had put it already in 1967; "Cyclic and Reciprocal Sedimentation," 810. For the extent of Gondwanaland glaciation, see Thomas J. Crowley and Steven K. Baum, "Estimating Carboniferous Sea-Level Fluctuations from Gondwanan Ice Extent," *Geology* 19 (October 1991): 975–77. For the links among glaciation, rainfall, and stratification, see C. Blaine Cecil, "Paleoclimate Controls on Stratigraphic Repetition of Chemical and Siliciclastic Rocks," *Geology* 13 (June 1990): 533–36.

18. Heckel, "Glacial-Eustatic Sea-Level Curve," 348.

19. Ibid. 355.

20. Heckel, "Evaluation of Evidence of Glacio-Eustatic Control," 81.

21. The Milankovitch cycles are treated in Ruddiman, *Earth's Climate*, chapter 8; and in Broecker, *The Glacial World According to Wally,* "Records" section, 12–24.

22. For ice packs to thicken, the summers have to be cold enough so that snow that fell in the Arctic winter stays frozen and gets compacted in the summer; interview, October 8, 2003.

23. Winchester, *The Map That Changed the World,* 47, 174–75.

24. Freese, *Coal*, 111–12.

25. Heckel interview.

26. Emerson is quoted in Freese, *Coal*, 10.

27. Although burning wood emits more carbon dioxide per unit energy than combustion of any other fuel, if new trees are planted immediately, all the carbon dioxide emitted is drawn back out of the atmosphere by their respiration as they grow. Accordingly, a wood economy kept in a steady state is carbon-neutral, but one in which wood is being depleted results in large net carbon dioxide emissions.

3. THE AIR WE BREATHE:
THE HUMAN COSTS OF COAL COMBUSTION

1. Interview with the author, March 26, 2004.

2. Freese, *Coal*, 40.

3. See Lomborg, *The Skeptical Environmentalist*, 163.

4. Ibid. 164.

5. Freese, *Coal*, 145–48.

6. Davis, *When Smoke Ran Like Water*, 45.

7. Ibid. 20.

8. The work by Lave and Seskin, along with Tukey's critique, is discussed by Davis in *When Smoke Ran Like Water*, 111–15. Lave confirmed in an interview, January 14, 2005, that Davis's account is accurate, if somewhat impassioned.

9. Lester B. Lave and Eugene P. Seskin, "Air Pollution and Human Health," *Science* 169 (August 21, 1970): 723–33.

10. Baltimore: Johns Hopkins University Press, 1977.

11. Davis, *When Smoke Ran Like Water*, 87.

12. Clean Air Task Force, *Death, Disease, and Dirty Power: Mortality and Health Damage Due to Air Pollution from Power Plants* (Boston, October 2000), 12–13, 3.

13. Because of a curious discrepancy between the figure in the body of the report and a number mentioned more casually in its preface by Dr. John Spengler of the Harvard School of Public Health, the report left some ambiguity whether it meant to put the figure at 30,000 or 3,000. But Spengler confirmed in a communication with the author, June 11, 2004, that the correct number is indeed in the tens of thousands, not thousands. Independently, in an interview on March 10, 2005, Dr. Jonathan M. Samet of the Bloomberg School of Public Health at Johns Hopkins University expressed confidence in the "transparency" of Spengler's work and in the soundness of his general conclusions.

14. U.S. Environmental Protection Agency (EPA), *The Benefits and Costs of the Clean Air Act, 1970 to 1990*, EPA 410-R-97-002, October 15, 1997, Washington, DC, 15.

15. Department of Energy, Energy Information Administration, *Effects of Title IV of the Clean Air Act Amendments of 1990 on Electric Utilities: An Update*, DOE/ EIA-0583 (97), March 1997, Washington, DC, executive summary, 2.

16. EPA and U.S. Department of State, *Climate Action Report*, 2002, chapter 3, Greenhouse Gas Inventory, Washington, DC, May 2002; *The Wall Street Journal*, November 8, 2004, A2.

17. See for example Kirk Johnson, "Coal Market Zooms in on Mines in Colorado," *The New York Times*, June 16, 2004, and Simon Romero, "Fuel of the Future," *The New York Times*, November 20, 2004; "Up Front," *BusinessWeek*, January 17, 2005, 11; Rebecca Smith, "After Taking Its Lumps, Coal Is Suddenly Hot Again," *The Wall Street Journal*, April 1, 2004; and "Global Surge in Use of Coal," *The Wall Street Journal*, November 16, 2004.

18. EPA, *Climate Action Report*, p. 37.

19. Bruce Barcott, "Changing All the Rules: How the Bush Administration Quietly—and Radically—Transformed the Nation's Clean Air Policy," *New York Times Magazine*, April 4, 2004, 73.

20. Greg Hitt, "Mercury Issue Takes Wing," *The Wall Street Journal*, June 9, 2004, A4. Hitt noted that a leading representative for industry was known in lobbying circles as "Mr. Mercury," and that Bush's close relationship to the coal industry dated to "2000, when coal producers delivered West Virginia, a [traditionally Democratic] battleground state crucial to putting it over the top." (Without West Virginia, the Florida recount would have been "just a footnote to a Gore presidency," observed Hitt.)

21. Clean Air Task Force, *Mercury and Midwest Power Plants*, Boston, MA, February 2003, 2; Allan Kolker, U.S. Geological Survey, talk at Lamont-Doherty Earth Observatory, May 28, 2004.

22. Phillip Babich, "Dirty Business: How Bush and His Coal Industry Cronies Are Covering Up One of the Worst Environmental Disasters in U.S. History," *Salon*, November 13, 2003 (http://www.salon.com/tech/feature/2003/11/13/slurry_coverup/index_np.html).

23. Francis C. Clines, "Judge Takes on the White House on Mountaintop Mining," *The New York Times*, May 19, 2002, 18. See also Michael Lipton, "The Fight for the Soul of Coal Country," *The New York Times*, May 17, 2002, Op-Ed page. For a radically depressing treatment of the subject, see Erik Reece, "Death of a Mountain: Radical Strip Mining and the Leveling of Appalachia," *Harper's* 311 (1859) (April 2005): 41–60.

24. Smil, *Energy at the Crossroads*, 233.

25. Doug Gibson of the UMWA provided the miner statistics, based on data from the National Mining Association, the Mine Safety Health Administration, and the union's own numbers.

26. William Sweet, "Nuclear Incident Could Prompt Costly Inspections," *IEEE Spectrum* 39 (6) (June 2002).

27. U.S. Canada Power System Outage Task Force, "Interim Report: Causes of the August 14th Blackout," November 19, 2003, https://reports.energy.gov/, 31, 33.

28. "Utilities and regulators are recognizing that it is unlikely that greenhouse gas emissions will continue to cost utilities nothing whatever over the long lifetime of new investments," and the "risk" of carbon regulation is therefore being incorporated quantitatively into long-term utility plans, says a recent article in a scholarly trade journal. See Karl Bokenkamp et al., "Hedging Carbon Risk: Protecting Customers and Shareholders from the Financial Risk Associated with Carbon Dioxide Emissions," *Electricity Journal* 18 (6) (July 2005): 11–23.

It's also been adduced, however, that a comparative advantage relative to utilities that burn western coal may play a role in their thinking. Appalachian coal is much higher in sulfur than western coal, which, since it consists to a great extent of lignite, is much less energy-efficient and much more carbon-intensive. That is, per kilowatt generated from western coal, significantly more carbon is emitted. See the Environmental Integrity Project's May 2005 report, "Dirty Kilowatts: America's Most Polluting Power Plants," tables 2 and 4. A .pdf can be accessed at http://www.environmentalintegrity.org/pub315.cfm.

4. FROM OUTER SPACE:

ASIA'S BROWN CLOUD, AND MORE

1. National Academy of Engineering, October 9, 2001.

2. John J. Fialka, "A Dirty Discovery Over Indian Ocean Sets Off a Fight," *The Wall Street Journal*, May 6, 2003; personal communications from Ramanathan.

3. See biography of Veerabhadran Ramanathan, *Proceedings of the National Academy of Sciences* 102 (15) (April 12, 2005): 5323–25.

4. V. Ramanathan, Bruce R. Barkstrom, and Edwin F. Harrison, "Climate and the Earth's Radiation Budget," *Physics Today* 42 (5) (May 1989): 22–32.

5. Ramanathan, interview with author, August 5, 2004.

6. Strictly speaking, the A stands for "afternoon." See Angelita C. Kelly and Edward J. Macie, "The A-Train: NASA's Earth Observing System (EOS) Satellites and Other Earth Observation Satellites," www.dlr.de/iaa.symp/archive/PDF_Files/IAA-B4–1507P.pdf, 1138–41. Some initial findings from the Terra and Aqua satellites, reported in a *Nature* press release as this book went to press, indicate that cooling effects from aerosols are at the high end of IPCC estimates. See article by Nicholas Bellouin and Jim Coakley, *Nature* 438 (7071) (December 22, 2005).

7. Ramanathan, interview with author, and *Proceedings of the National Academy of Sciences* biography.

8. Personal communication from Crutzen, April 19, 2005.

9. See the 2002 report to the United Nations Environment Program (UNEP), prepared by the Center for Clouds, Chemistry and the Climate (C4), "The Atmospheric Brown Cloud: Climate and other Environmental Impacts," http://www.rrcap.unep.org/issues/air/impactstudy/index.cfm.

10. V. Ramanathan, C. Chung, D. Kim, T. Beltge, L. Buja, J. T. Kiehl, W. M. Washington, Q. Fu, D. R. Sikka, and M. Wild, "Atmospheric Brown Clouds: Impacts on South Asian Climate and Hydrological Cycle," *Proceedings of the National Academy of Sciences* 102 (15) (April 12, 2005): 5326–33.

11. Fialka, "A Dirty Discovery."

12. Take note in particular of the furious debate over Bjørn Lomborg's *The Skeptical Environmentalist: Measuring the Real State of the World* (Cambridge University Press, 1998), which is easily tapped into online.

13. H. Keith Florig, "China's Air Pollution Risks," *Environmental Science & Technology* 31 (6): 274–79. In an interview with the author on August 5, 2004, Florig stood by his conclusions and considered them by and large still current.

14. Cited in Ted C. Fishman, *China, Inc.* (New York: Scribner, 2005), 116. China's cement production, together with its coal combustion, also is a major source of mercury, raising atmospheric levels all over the world; see Matt Pottinger, Steve Stecklow, and John J. Fialka, "A Hidden Cost of China's Growth," *The Wall Street Journal*, December 17, 2004, 1.

15. Vaclav Smil, "China Shoulders the Cost of Environmental Change," *Environment* 39 (6) (July–August 1997): 9.

16. Lomborg, *The Skeptical Environmentalist*, 175.

17. See The World Bank, *Clear Water, Blue Skies: China's Environment in the New Century* (Washington, DC: The World Bank, 1997), 18.

18. Kirk R. Smith, "National Burden of Disease in India from Indoor Air Pollution," *Publications of the National Academy of Sciences* 97 (24) (November 21, 2000): 13286–93.

19. Interview with author, January 20, 2005.

20. Smith questions the alleged carbon-neutrality of biomass in a self-replenishing biomass economy. Even if regeneration of crops is 100 percent, when combustion inefficiencies and all the by-products are taken into account, a switch to natural gas may result in lower greenhouse gas emissions per unit of energy produced and consumed. If biogas is generated from properly regenerated agricultural resources, it appears on balance to be the best replacement for home burning of sticks, dung, or charcoal briquettes. Natural gas is the next best and in urban areas the only practicable option. See Rufus D. Edwards, Kirk R. Smith, Junfeng Zhang, and Yuqing Ma, "Implications of Changes in Household Stoves and Fuel Use in China," *Energy Policy* 32 (2004): 395–411. See also Kirk R. Smith, Rufus D. Edwards, Junfeng Zhang, and Yuqing Ma, "Greenhouse Implications of Household Stoves: An Analysis for India," *Annual Review of Energy and the Environment* 25 (2000): 741–63.

21. Marlowe Hood and William Sweet, "Energy Policy and Politics in China," *IEEE Spectrum* 35 (11) (November 1999): 36–37.

22. Ibid.

23. Fishman, *China, Inc.,* 113.

24. See William Sweet, "China's Big Dams," *IEEE Spectrum* 37 (9) (September 2001). Rotting vegetation in reservoirs behind dams may be a substantial source of methane, a potent greenhouse gas. So ranking hydroelectricity with nuclear electricity as "zero carbon" may not be justified.

25. China displaced Japan as the world's second leading petroleum importer in 2003–4 (behind the United States); its appetite for oil was considered a major factor in the sharp escalation of oil prices in 2004.

26. Mining accidents routinely claim hundreds of lives in today's China—see Stephanie Hou, "Blast in Chinese Mine Kills 209," *USA Today*, February 16, 2005, 8a.

27. World Bank reports: *Clear Water, Blue Skies: China's Environment in the New Century* (Washington, DC: The World Bank, 1997), 54; Todd M. Johnson et al., *China: Issues and Options in Greenhouse Gas Emissions Control*, discussion paper no. 330 (Washington, DC: The World Bank, 1996), 62.

PART 2. CLIMATE: THE LOCKSTEP RELATIONSHIP
BETWEEN CARBON DIOXIDE AND TEMPERATURE

5. THE DRILLERS

1. Fagan, *The Little Ice Age*, xv.

2. Langway considers Bader the real father of ice core research. Richard B. Alley says that while nobody knows who first started noticing the intriguing layers in the walls of Alpine glaciers, many observers indeed credit Bader with "initiating the modern era of ice-core drilling that ultimately led us to Greenland." See Alley, *The Two-Mile Time Machine*, 17.

3. C. C. Langway Jr. and B. Lyle Hansen, "Drilling Through the Ice Cap: Probing Climate for a Thousand Centuries," *The Bulletin of the Atomic Scientists* 26 (10) (December 1970): 63. Unless otherwise indicated, all background on Langway and his work with colleagues is from interviews conducted on Cape Cod on January 17, 2005, and by telephone on December 8, 2004.

4. Henri Bader, "United States Polar Ice and Snow Studies in the International Geophysical Year," *Proceedings of the Symposium at the Opening of the International Geophysical Year* (June 28–29, 1957) (Washington, DC: U.S. National Committee for the IGY, 1958).

5. C. C. Langway and Hans Oeschger, "Introduction: Glaciers as a Source of Earth History," in *The Environmental Record in Glaciers and Ice Sheets*, ed. C. C. Langway and Hans Oeschger (New York: Wiley, 1989), 5.

6. The French mathematician and physicist Joseph Fourier is credited with having recognized the greenhouse effect, though he somewhat misconceived it. The Englishman John Tyndell clinched the theory in 1862 with his discovery that water and carbon dioxide are opaque to heat radiation.

7. The Swede Svante Arrhenius recognized that rising carbon dioxide levels would warm the earth's atmosphere, and he did the first estimate of how much warming a doubling of carbon dioxide would induce. His result was remarkably accurate, by current reckoning, though his method was faulty, as discussed in chapter 6.

8. Fagan, *The Little Ice Age*, 52, 54.

9. A noted report on the 1,000-year temperature record is Michael Mann et al., "Northern Hemisphere Temperatures During the Last Millennium," *Geophysical Research Letters* 26 (March 15, 1999): 759–62. What's come to be known as Mann's hockey stick curve has come under sharp criticism but, in this writer's view, has largely stood up to scrutiny. As this book went to press, T. J. Osborn and K. R. Briffa reported in the February 10, 2006 issue of *Science* that twentieth-century warming is the most widespread and the longest temperature anomaly since the ninth century—as far back as observations support such an analysis.

10. Dansgaard, *Frozen Annals*, 11.

11. Richard Rhodes, *The Making of the Atomic Bomb* (New York: Simon and Schuster, 1988), 759–62.

12. Harold Urey already had discovered in 1947 that they could be measured in the shells of prehistoric foraminifera ("forams"), found in ocean floor borings, to

estimate the temperature of the ocean at the time they lived, according to Spencer Weart, *The Discovery of Global Warming,* 46.

13. Dansgaard, *Frozen Annals,* 13.

14. The IAEA was established in the mid-1950s to safeguard nuclear technology provided by nuclear weapons-producing countries to non-nuclear customers in bilateral deals. With the conclusion of the Nuclear Nonproliferation Treaty in the 1970s, its responsibilities expanded enormously to cover, eventually, virtually all nuclear equipment operated and materials used in non-nuclear member states.

15. Willi Dansgaard, "Stable Isotopes in Precipitation," *Tellus* 16 (1964): 436–68.

16. See Willi Dansgaard, "The O-18 Abundance in Fresh Water," *Geochimica et Cosmochimica Acta* 6 (5–6) (1954): 241–60.

17. Dansgaard, *Frozen Annals,* 16. In the words of Richard B. Alley, an ice core specialist at Pennsylvania State University and the leader of a recent National Academy survey of abrupt climate change, Dansgaard's insights "clearly [opened] the possibility of paleochemistry," that is, the study of the ancient earth's atmospheric compositions. "Old water [could] be found in the ice of an ice sheet or in well water, or in water that was taken up by a tree as it grew and was used to make wood. Shell-building animals [in the ocean] take oxygen from water and store it in their shells, so old shells record the composition of the water. This is a powerful technique, and it would be used routinely to learn past temperature" (*The Two-Mile Time Machine* 64).

18. Denmark had given the United States the right to make military use of Greenland in 1941 but in 1947 rescinded that right. After further negotiations, a 1951 agreement allowed the United States to build an air force base at Thule and NATO members to make use of all military facilities on the island. See John D. Cox, *Climate Crash: Abrupt Climate Change and What It Means for Our Future* (Washington, DC: Joseph Henry Press, 2005), 34.

19. Ibid. 39–41.

20. Sigfus Johnsen, interview with the author, November 30, 2004.

21. See obituary by Thomas Stocker, *Nature* 397 (6718) (February 4, 1999): 396. Oeschger appears to have taken a stronger interest in the social implications of scientific discoveries than Dansgaard, and considered it the responsibility of every scientist to bring important results to the public's attention. "The worst for me would be if there were serious changes in the next five to ten years [in the earth's climate] and we scientists … did not have the courage to point out these dangerous developments early on," he once said, according to the obit.

22. Dansgaard, *Frozen Annals,* p. 57.

23. Dansgaard et al., "One Thousand Centuries of Climate Records from Camp Century on the Greenland Ice Sheet," *Science* 166 (1969): 377–81.

24. DEW stood for Distant Early Warning (of Soviet missile attack).

25. Dansgaard, *Frozen Annals,* p. 69.

26. According to Weart, *The Discovery of Global Warming,* 74.

27. For the account of Dansgaard–Oeschger and Heinrich–Bond cycles that follows, see Alley, *The Two-Mile Time Machine,* 154–58.

28. See William F. Ruddiman and A. McIntyre, "The North Atlantic Ocean During the Last Deglaciation," *Paleogeography, Paleoclimatology, and Paleoecology* 35 (1981): 145–214.

29. Weart, *The Discovery of Global Warming*, 74.

30. Ibid. 47.

31. Elizabeth Kolbert, "Ice Memory," *The New Yorker*, July 1, 2002.

32. Fagan, *The Little Ice Age*, 66.

33. Cox, *Climate Crash*, 77.

34. Werner Berner, Hans Oeschger, and Bernhard Stauffer, "Information on the CO_2 Cycle from Ice Core Studies," *Radiocarbon* 22 (2) (1980): 227–35.

35. A. Neftel, Hans Oeschger, et al., "Ice Core Sample Measurements," *Nature* 295 (5846) (January 21, 1982): 220–23.

36. A. Neftel et al., "CO_2 Record in the Byrd Ice Core," *Nature* 331 (6157) (February 18, 1988): 609–11.

37. Quoted in Cox, *Climate Crash*, 99.

38. A. Neftel et al., "Evidence from Polar Ice Cores for the Increase in Atmospheric CO_2 in the Past Two Centuries," *Nature* 315 (6014) (May 2, 1985): 45–47.

39. Eric T. Sundquist, "Ice Core Links CO_2 to Climate," *Nature* 329 (6138) (October 1, 1987): 389–90.

40. The Stauffer commentary appeared on pages 412–13, and the article by J. R. Petit et al., "Climate and Atmospheric History of the Past 420,000 Years from the Vostok Ice Core, Antarctica," on pages 429–35 of *Nature* 399 (6735) (June 3, 1999).

41. Other recipients of the Tyler Prize include the population experts Anne and Paul Ehrlich, and Joel Cohen; the naturalists Jane Goodall and George Schaller; and the climate scientists Broecker and—most recently—Keeling, mainly for his Mauna Loa carbon dioxide measurements, and Lonnie Thompson, the world's foremost expert on tropical mountain glaciers.

42. "Eight Glacial Cycles from an Antarctic Ice Core," by European Project for Ice Coring in Antarctica (EPICA) community members, *Nature* 429 (June 10, 2004): 623–28. It's important to note that the linear relationship between greenhouse gases and temperature has been demonstrated only for the last 700 millennia; there's substantial evidence that it may not hold for some earlier eras such as the Miocene, 25 to 9 million years ago. The crucial point, nonetheless, is that the relationship has been proved almost beyond doubt for the world in which we live today, the world of the past 650,000 to 700,000 years. As this book went to press, Thomas Stocker, Oeschger's successor in Bern, reported with colleagues on their analysis of the latest ice core results from Antarctica: in a press advisory issued by *Science* magazine on November 21, 2005, the new work, they said, "confirms the stable relationship between Antarctic climate and carbon dioxide and methane during the last four glacial cycles and extends this stable relationship back another two glacial cycles."

43. For relatively recent discussions of the leads and lags, see Wallace S. Broecker, "Does the Trigger for Abrupt Climate Change Reside in the Ocean or in the Atmosphere?" *Science* 300 (June 6, 2003): 1519–22, and Richard A. Kerr,

"Ocean Flow Amplified, Not Triggered, Climate Change," *Science* 307 (March 25, 2005): 1854.

44. Dansgaard, *Frozen Annals*, 119.

6. THE MODELERS

1. For Lorenz's discovery of the butterfly effect, see the whole first chapter of Gleick, *Chaos*.

2. Observations of Manabe, and about him, are from personal observation at the San Diego symposium and banquet, from remarks made there by friends and colleagues, and from a long interview conducted with him at his home on January 27, 2005.

3. Jule G. Charney, R. Fjortoft, and John von Neumann, "Numerical Integration of Barotropic Vorticity Equation," *Tellus* 2 (1950): 237–54.

4. For background on von Neumann and the ENIAC, see Ceruzzi, *A History of Modern Computing*, 19–24.

5. Quotes in Sheets and Williams, *Hurricane Watch*, 203.

6. Three of the so-called primitive equations are derived from Newton's laws of motion and two from thermodynamics. A sixth—a continuity equation—accounts for columns of air entering or leaving a defined area. For a basic account, see ibid. 204–5.

7. American Institute of Physics, History Center, "General Circulation Model Historical Overview," www.aip.org/history/cloan/gcm/histoverview.html.

8. At the end of 2004 the Earth Simulator was superseded by even more capable machines from IBM and SGI, and it was clear that advances in supercomputer performance would continue to accelerate. See Erico Guizzo, "IBM Reclaims Supercomputer Lead," *IEEE Spectrum* 42 (2) (February 2005).

9. All statements directly attributed to Manabe are from the four-hour interview conducted at his home in January 2005 or from the San Diego symposium and banquet on January 12.

10. Weart, *The Discovery of Global Warming*, 107.

11. For background on Arrhenius, see ibid. 5–7.

12. Interview, January 27, and reference 19 in "Simple Models of Climate," American Institute of Physics, History Center, www.aip.org/history/climate.

13. Manabe, "Early Development in the Study of Greenhouse Warming: The Emergence of Climate Models," *Ambio* 26 (1) (February 1997): 47–51.

14. A complete list of Manabe's publications with all bibliographical information can be found at http://www.gfdl.noaa.gov/reference/bibliography/authors/manabe.html.

15. Communication with author, May 27, 2005.

16. Interview with author, January 24, 2005.

17. Quoted in Paul Edwards and Spencer Weart, "General Circulation Models," American Institute of Physics, History Center, 18.

18. S. Manabe, K. Bryan, and M. J. Spelman, "A Global Ocean-Atmosphere Climate Model with Seasonal Variation for Future Studies of Climate Sensitivity," *Dynamics of Atmospheres and Oceans* 3 (1979): 395–96.

19. For background on spherical harmonics, see Edwards and Weart, "General Circulation Models."

20. San Diego symposium: talk by Keith W. Dixon.

21. Sydney Levitus, interview with author, January 26, 2005.

22. Sydney Levitus and others recently estimated that from 1955 to 1998, the world ocean heat content increased by 14.5 x 10^{22} joules, corresponding to a mean temperature increase of $0.037°C$, at a rate of 0.20 watts per square meter of the earth's total surface area. See "Warming of the World Ocean, 1955–2003," *Geophysical Research Letters* 32 (January 22, 2005), L02604, doi [document object identifier] 10.1029/2004GL021592.

23. Timothy P. Boyer, Sydney Levitus, et al., "Linear Trends in Salinity for the World Ocean, 1955–1998," *Geophysical Research Letters* 32 (January 6, 2005), L01604, doi 10.1029/2004GL021791.

24. William Sweet, "British Meteorological Office Spins Off Climate Research Center," *Physics Today* 43 (6) (June 1990): 67.

25. Weart, *The Discovery of Global Warming*, 133, 170–71; William Sweet, "Power and Energy," *IEEE Spectrum* 32 (1) (January 1995).

26. See Robert J. Charlson and Tom M. L. Wigley, "Sulfate Aerosol and Climatic Change," *Scientific American* 282 (2) (February 1994): 48–57. That warming in the twentieth century was not as great as the basic greenhouse gas model would have predicted, and the irregular pattern of warming in the last century, have figured prominently in skeptical views of global warming. See for example Jack M. Hollander, "Rushing to Judgment," *The Wilson Quarterly* 27 (2) (Spring 2003): 67–69.

27. Charlson and Wigley, "Sulfate Aerosol and Climate Change," 56.

28. Qiang Fu et al., "Contribution of Stratospheric Cooling to Satellite-Inferred Tropospheric Temperature Trends," *Nature* 429 (6982) (May 6, 2004): 55–58.

29. A key subcommittee is chaired by John Christy of the National Climate Data Center, and the executive summary will be written by NCAR's Tom Wigley, a coauthor of the *Scientific American* article cited above.

30. Peter Foukal, Gerald North, and Tom Wigley, "A Stellar View on Solar Variations and Climate," *Science* 306 (October 1, 2004): 69.

31. From various geological clues, it now seems that this nightmarish scenario may actually have occurred 600 to 700 million years ago, though the earth was not completely lifeless.

32. American Institute of Physics, History Center, "General Circulation Models," 23.

33. In 1995 and 1997, Thompson and colleagues published results from two cores drilled on Huascaran, 9 degrees south of the equator in the Andes, showing the same isotopic depletion seen in the polar cores. A core from Mount Sajama in Bolivia showed that there were "climate shifts in the tropics of the same magnitude that Greenland experienced during the ice ages," as Thompson put it to Virginia Morell in "Time Signs," *National Geographic* (September 2004): 72. According to a personal communication on February 2, 2005, Thompson's group is drilling a

new core at Peru's Coropuna, analysis of which they believe will push the tropical temperature record back 22,000 years.

34. See for example S. Manabe and R. J. Stouffer, "Simulation of Abrupt Climate Change Induced by Freshwater Input to the North Atlantic Ocean," *Nature* 378 (November 9, 1995): 165–67.

35. Manabe et al., "Century-Scale Change in Water Availability: CO_2-Quadrupling Experiment," *Climatic Change* 64 (1–2): 59–76.

36. See A. Berger and M.-F. Loutre, "Paleoclimate Sensitivity to CO_2 and Insolation," *Ambio* 24 (1) (1997): 32–37.

37. Q&A by Claudia Dreifus, *The New York Times*, December 16, 2003.

7. THE SYNTHESIZERS

1. The establishment of the Stanford institute was reported in *The Wall Street Journal* and *The New York Times* on November 21, 2002. Schneider delivered one such talk at a workshop sponsored by the National Association of Science Writers in Denver, Colorado, on February 13, 2003.

2. For profiles and discussions of Sachs, see John Cassidy's review in *The New Yorker*, April 11, 2005, 72–77, and John H. Richardson's short article in *Esquire*, December 2003, 197–200 and 212.

3. The author conducted a full day of interviews about the Earth Institute at the Lamont-Doherty Earth Observatory on January 17, 2003; much of the background on the institute's origins came from Peter Schlosser, the chairman of its academic committee, and from a follow-up interview on January 28, 2003, with its deputy director, John C. Mutter. Michael Crowe, its founding director, provided some additional information in an informal conversation on April 22, 2003.

4. According to Hallam, *Great Geological Controversies*, 144, the Lamont-Doherty lab "had been built up in the post-war years by its dynamic director Maurice Ewing into one of the greatest institutes of geological and geophysical oceanography in the world. The institute's general bias towards data-collecting on a massive scale, and hostility toward mobilistic views [associated with the theory of continental drift] reflected the outlook and attitude of its director." Many of its staff were converted to the theory of plate tectonics only in the 1960s, when results came in from a study of magnetic anomalies across the Pacific–Antarctic ridge and provided decisive evidence in favor of the theory.

5. The author conducted a formal, taped two-hour interview with Sachs on January 28, 2003.

6. Biographical detail on Hansen is from interviews conducted on January 2 and March 21, 2003; see also the short profile by David Goeller in *Environmental Action* 20 (6) (November–December 1989): 24–25.

7. Quoted in Christianson, *Greenhouse*, 171.

8. J. E. Hansen et al., "The Climate Impact of Increasing Carbon Dioxide," *Science*, vol. 213, pp. 957–966.

9. See http://www.columbia.edu/library/earthmatters/spring2000/pages/page9.html.

10. J. E. Hansen et al., "Global Warming in the Twenty-First Century: An Alternative Scenario," *Proceedings of the National Academy of Sciences* 97 (2000): 9875–80.

11. "Can We Defuse the Global Warming Time Bomb?" based on Hansen's presentation to the Council on Environmental Quality, June 1, 2003, can be accessed at http://pubs.giss/nasa/gov/authors/jhansen/.

12. For the Faustian bargain, see J. E. Hansen and Andrew A. Lacis, "Sun and Dust Versus Greenhouse Gases," *Nature* 346 (6286) (August 23, 1990): 713–19.

13. Walter Robinson, professor of atmospheric sciences, University of Illinois, Urbana-Champaign, communication with the author.

14. www.scienceexpress.org, 1110252.

15. See Wallace S. Broecker's autobiographical sketch, "Converging Paths Leading to the Role of the Oceans in Climate Change," *Annual Review of Energy and the Environment*, 25 (2000): 19. See also the oral history interview available by searching the International Catalog of Sources at the American Institute of Physics Web site on history of climate change at http://www.aip.org/history/.

16. "Unpleasant Surprises in the Greenhouse," *Nature* 328 (6126) (July 9, 1987): 123–27; the Biggest Chill," *Natural History* 96 (10) (October 1987): 74–82; Broecker quote is from his autobiographical sketch.

17. For a summary of the National Academy report, see R. B. Alley et al., "Abrupt Climate Change," *Science* 299 (March 28, 2003): 2005–10. The full report is *Abrupt Climate Change: Inevitable Surprises* (Washington, DC: National Academy Press, 2002). It can be accessed online at http://books.nap.edu/books/0309074347/html/1.html#pagetop.

18. The Pentagon report, "An Abrupt Climate Change Scenario and Its Implications for United States National Security," written by Peter Schwartz and Doug Randall and completed in October 2003, was initiated by a famous veteran intelligence guru at the Pentagon, Andrew W. Marshall, who has been a confidante of Defense Secretary Donald Rumsfeld for many decades. For more temperate treatments of the same subject, see Edouard Bard, "Climate Shock: Abrupt Changes Over Millennial Time Scales," *Physics Today* 55 (12) (December 2002): 32–38, and Spencer Weart, "The Discovery of Rapid Climate Change," *Physics Today* 56 (8) (August 2003): 30–36.

19. Mark Z. Jacobson, "Control of Fossil-Fuel Particulate Black Carbon and Organic Matter, Possibly the Most Effective Method of Slowing Global Warming," *Journal of Geophysical Research* 107, D19 (2002): 4410–32. Jacobson's description of the White House reaction is at his Web site, http://www.stanford.edu/group/efmh/bush/.

PART 3. CHOICES: THE LOW-CARBON AND ZERO-CARBON TECHNOLOGIES WE CAN DEPLOY RIGHT NOW

8. BREAKING THE CARBON HABIT

1. One researcher looked at 928 technical papers on climate published between 1993 and 2003 and found, "remarkably," that "none disagreed with the consensus

position." See Naomi Oreskes, "The Scientific Consensus on Climate Change," *Science* 306 (December 3, 2004): 1686.

2. See Maxwell T. Boykoo and July M. Boykoff, "Balance as Bias: Global Warming and the U.S. Prestige Press," *Global Environmental Change* 14 (2004): 125–36.

3. For example, one survey found that when scientists were asked what the chances of an "unlikely" event happening actually were, responses ranged from 0 to 40 percent; the result is cited in M. Granger Morgan and David W. Keith, "Subjective Judgments by Climate Experts," *Environmental Science and Technology* 29 (1995): 468–76. The issue is further complicated by the fact that scientists themselves influence the probability or improbability of future scenarios; see the editorial comments "Incorporating Agency Into Climate Change Risk Assessments" and "Agency and the Assignment of Probabilities to Greenhouse Emissions Scenarios," *Climatic Change* 67 (1) (November 2004): 13–42.

4. An expert conference on dangerous climate change was convened in Exeter, UK, February 1–3, 2005. See http://www.stabilisation2005.com/. For attempts at a rigorous treatment of dangerous climate change, which can get quite arcane, see Michael D. Mastrandea and Stephen H. Schneider, "Probabilistic Integrated Assessment of 'Dangerous' Climate Change," *Science* 304 (April 23, 2004): 571–75; and Tom M. L. Wigley, "Choosing a Stabilization Target for CO$_2$," *Climatic Change* 67 (1) (November 2004): 1–11.

5. William D. Nordhaus, "Global Warming Economics," *Science* 294 (November 9, 2001): 1283–84.

6. Donald Kennedy, "Climate and Civilization: The Scientific Evidence for Climate Change, and How Our Response to It May Influence National Policy," lecture on the science of climate warming delivered at the State University of New York at Oneonta, October 4, 2000; accessible at www.oneonta.edu/faculty/allenth/climate_and_civilization.

7. Alley talk, session on detection and attribution of climate trends, American Association for the Advancement of Science (AAAS) annual meeting, Washington, DC, February 18, 2005.

8. Larry Rohter, "Antarctica, Warming, Looks Ever More Vulnerable," *The New York Times*, January 25, 2005, science section lead story.

9. Alley talk, session on detection and attribution of climate trends, American Association for the Advancement of Science (AAAS) annual meeting, Washington, DC, February 18, 2005.

10. Evan Mills of Lawrence Berkeley National Laboratory, talk at climate change and insurance session, AAAS annual meeting, Washington, DC, February 18, 2005.

11. Talk by Christoph Schär at climate modeling session, American Meteorological Society (AMS) annual meeting, San Diego, California, January 11, 2005. A different group, however, estimates that the heat wave was merely the worst since the year 1500: see Peter A. Stott, D. A. Stone, and M. R. Allen, "Human Contribution to the European Heat Wave of 2003," *Nature* 432 (December 2, 2004): 610–14.

12. Suki Manabe Symposium, American Meteorological Society annual meeting, San Diego, California, January 10, 2005.

13. L. J. Mickley, D. J. Jacob, and B. D. Field, "Effects of Future Climate Change on Regional Air Pollution Episodes in the United States," *Geophysical Research Letters* 31, L24103, doi: 10.1029/2004 GL 02121b, 2004.

14. See *National Geographic*'s special report, "Global Warming: Bulletins from a Warmer World" 26 (2) (September 2004); Richard B. Alley, "Abrupt Climate Change," *Scientific American* 293 (4) (November 2004): 62–69; and Edouard Bard, "Climate Shock: Abrupt Changes Over Millennial Time Scales," *Physics Today* 55 (12) (December 2002): 32–38.

15. Although the widely noted recent book by the geographer Jared Diamond, *Collapse: How Societies Choose to Fail or Succeed* (Viking, 2004), is not explicitly about contemporary climate change, the civilization-threatening aspect of global warming is a latent theme.

16. Don Wuebbles, talk at climate and insurance session, AAAS annual meeting, Washington, DC, February 18, 2005.

17. Quoted by Evan Mills, insurance session, AAAS annual meeting, Washington, DC, February 18, 2005.

18. See Posner, *Catastrophe*, especially 43–58 and 253–60. Discussion of the interesting and important issues raised in his book can be followed at www.becker-posner-blog.com, a site that he shares with University of Chicago economist Gary Becker.

19. Richard A. Posner, "The Probability of Catastrophe," *The Wall Street Journal*, January 6, 2005.

20. *Leo Szilard: His Version of the Facts*, ed. Spencer R. Weart and Gertrud Weiss Szilard (Cambridge: MIT Press, 1978), 54.

21. See Doris Kearns Goodwin, *No Ordinary Time: Franklin and Eleanor Roosevelt, the Home Front in World War II* (New York: Simon and Schuster, 1994), 194.

22. See National Research Council, *Novel Approaches to Carbon Management*, 13. Another way to look at the quantities is explained in the longer online text that accompanied the summary article, S. Pacala and R. Socolow, "Stabilization Wedges: Solving the Climate Problem for the Next 50 Years with Current Technology," *Science* 305 (August 13, 2004): 968–72, and is discussed in more detail in chapter 9. Pacala and Socolow find that the flows of carbon dioxide into and out of natural gas storage reservoirs needed to achieve one of their "stabilization wedges" would be twenty times the current amount. A wedge would require about ten times the quantity currently produced in connection with commercial hydrogen production. In terms of storage, the amount of carbon dioxide sequestered would be 100 times the quantities now pumped into the ground in enhanced oil recovery operations. See the Supporting Online Material, www.sciencemag.org/cgi/content/full/305/5686/968/DC1, 17–19.

23. Interview with the author, June 18, 2001. For a fuller version of his views, see his "What Future for Carbon Capture and Sequestration?" *Environmental Science & Technology* 35 (7) (April 1, 2001): 148–53. Socolow also estimates the cost of carbon capture (not including storage) at 2 cents per kilowatt-hour, in "Can We Bury Global Warming?" *Scientific American* 294 (7) (July 2005): 52.

24. See Vanessa Houlder, "The Case for Carbon Capture and Storage," *The Financial Times*, January 23, 2004.

25. Herzog, "The Top Ten Things You Should Know About Carbon Sec tion," in National Research Council, *The Carbon Dioxide Dilemma*, 118.

26. For an excellent account of the BP project in North Africa, see the page article in *The Wall Street Journal*, February 4, 2005.

27. See *BusinessWeek*, "Putting Carbon Dioxide in Its Place" (October 20, 2003):82–83. John Davidson of the International Energy Agency in Paris has estimated that switching to fuels that are cheaper than or nearly as cheap as coal— notably natural gas and nuclear fuels—will be considerably less expensive than sequestration for at least another fifteen to twenty-five years. Electricity from wind turbines probably is a less expensive option too, where conditions are right, and electricity from biomass is about as good. Quoted in Houlder, "The Case for Carbon Capture and Storage."

28. "National Research Council, *Novel Approaches to Carbon Management*, 1.

29. Ibid. 14–15.

30. Ibid. 26.

31. Socolow, "The Century-Scale Problem of Carbon Management," in National Research Council, *The Carbon Dioxide Dilemma*, 13.

32. Orr, "Sequestration via Injection of Carbon Dioxide into the Deep Earth," in National Research Council, *The Carbon Dioxide Dilemma*, 21.

33. Hill, "Using Carbon Dioxide to Recover Natural Gas and Oil," in National Research Council, *The Carbon Dioxide Dilemma*, 23.

34. Benson, "Geological Sequestration of Carbon Dioxide," in National Research Council, *The Carbon Dioxide Dilemma*, 32–33.

35. Hill, "Using Carbon Dioxide to Recover Natural Gas and Oil," 26.

36. On February 10, 2006, when BB and Edison International announced plans to build a plant in Carson, California, in which carbon would be extracted from petroleum coke (precombustion) and sequestered in nearby oil fields owned by Occidental Petroleum, they said the cost of the plant per unit energy capacity would be twice that of wind turbines, which are merely near-commercial.

37. See John J. Fialka, "Bush Team Faces Serious Challenges on Mercury Rules," *The Wall Street Journal*, March 15, 2005.

38. George F. Kennan, *American Diplomacy: 1900–1950* (Chicago: University of Chicago Press, 1951), 73.

9. GOING ALL OUT FOR RENEWABLES, CONSERVATION, AND GREEN DESIGN

1. See Thomas R. Casten and Brennan Downes, "Critical Thinking About Energy: The Case for Decentralized Generation of Electricity," *Skeptical Inquirer* 29 (1) (January–February 2005): 26

2. Martin L. Hoffert et al., "Advanced Technology Paths to Global Climate Stability: Energy for a Greenhouse Planet," *Science* 298 (November 1, 2002): 981–87.

3. The solar satellite, first proposed in 1968, may in principle be feasible, but many technology breakthroughs would be required to make the concept work. Applications of superconducting technology in the power sector are very promising

in the medium term but have been disappointingly slow to mature in the short term. Since the discovery in the 1980s of materials that conduct electricity with zero resistance at liquid nitrogen temperatures, strenuous efforts have been made in all the advanced industrial countries to develop transmission cable, electricity generators, transformers, electric motors, and a variety of other specialized devices in which the new materials would be used. Steady advances are taking place, but none of those technologies has made a breakthrough to commercial viability yet, and so none can be considered a demonstrated, viable means of achieving reductions in carbon emissions right now. For a survey, see *IEEE Spectrum,* special issue on high-temperature superconductivity in power applications, 34 (7) (July 1997).

4. S. Pacala and R. Socolow, "Stabilization Wedges: Solving the Climate Problem for the Next 50 Years with Current Technology," *Science* 305 (August 13, 2004): 968–72.

5. Worldwatch Institute, *Vital Signs 2005* (Washington, DC: Worldwatch Institute, 2005), 14.

6. See Peter Fairley, "Gigawatts from Gusts," *Spectrum Online* (May 4, 2005), http://www.spectrum.ieee.org/WEBONLY/wonews/may05/0505nwind.html.

7. See William Sweet, "Wind Energy Takes Unfortunate Turn," *IEEE Spectrum* 41 (11) (November 2004) and Robb Mandelbaum, "Reap the Wild Wind," *IEEE Spectrum* 39 (10) (October 2002).

8. See William Sweet, "Energy Answer—Blowing in the Wind," *IEEE Spectrum* 41 (2) (February 2004).

9. See Peter Fairley, "The Greening of GE," *IEEE Spectrum* 42 (7) (July 2005).

10. Deutch and Lester, *Making Technology Work*, 35. Deutch was President Carter's director of energy research, and defense secretary in President Clinton's second term.

11. Smil, *Energy at the Crossroads*, 276.

12. See Peter Fairley, "Steady as She Blows," *IEEE Spectrum* 40 (8) (August 2003).

13. Smil, *Energy at the Crossroads*, 287.

14. Deutch and Lester, *Making Technology Work*, 43.

15. See John P. Benner and Lawrence Kazmerski, "Photovoltaics: Gaining Greater Visibility," *IEEE Spectrum* 36 (9) (September 1999).

16. Ibid.; Smil, *Energy at the Crossroads*, 287.

17. This was a refrain at the World Renewable Energy Congress VII held in Cologne, Germany, from June 29 to July 5, 2002.

18. Smil, *Energy at the Crossroads*, 284–85.

19. See Peter Fairley, "BP Solar Ditches Thin-Film Photovoltaics," *IEEE Spectrum* 40 (1) (January 2003). BP's posture, though not completely hollow, was based on its acquisition of a tiny U.S. photovoltaics manufacturer that happened to be at that time the world's biggest.

20. See Peter Fairley, "Solar Cell Rollout," *Technology Review* 107 (6) (July–August 2004): 35–40; Peter Fairley, "Can Organics Replace Silicon in PV?" *IEEE Spectrum* 41 (1) (January 2004).

21. Deutch and Lester, *Making Technology Work*, 43.

22. Ibid. 17.

23. Amory B. Lovins, "Energy Strategy: The Road Not Taken," *Foreign Affairs* 55 (1): 65–96.

24. Smil, *Energy at the Crossroads,* 166.

25. President Carter, incidentally, made Lovins the head of his newly established Solar Energy Institute in Golden, Colorado, which has evolved into the National Renewable Energy Laboratory.

26. Pacala and Socolow, "Stabilization Wedges," 969.

27. Huber and Mills, *The Bottomless Well,* 110.

28. *BusinessWeek* (February 14, 2005):67.

29. For all practical purposes the Cheney report gave up on the ideas of energy independence and energy security, calling instead for creative diplomacy to secure new sources of oil; from that point of view, the report was of a piece with the redeployment of U.S. military forces into gas-rich central Asia and the oil-rich Middle East, which began after 9/11. The *National Energy Report* was released on May 16, 2001 and is readily found on the Web.

30. *Clean Energy Scenarios,* coordinated by Marilyn A. Brown, Oak Ridge National Laboratory, Mark D. Levine, Lawrence Berkeley National Laboratory, and Walter Short, National Renewable Energy Laboratory, November 2000. For a closer comparison of this and the Cheney report, see *IEEE Spectrum* (July 2001).

31. Casten and Downes, "Critical Thinking About Energy," 30, figure 8.

32. The Public Utilities Regulatory Policy Act (PURPA), enacted during the Carter years, required utilities to purchase electricity from independent power producers and led to a substantial increase in cogeneration. The Energy Policy Act of 1992, though it had some unwelcome consequences as discussed in chapter 3, required the whole electricity system to be made open-access, giving a further boost to independent power producers deploying new technology.

33. See Gissen, ed., *Big and Green.* The book features designs by a number of architects noted for innovative green concepts, including Norman Foster and the Richard Rogers Partnership in London and, in New York City, Kiss + Cathcart, Cook + Fox, and Fox & Fowle, the designer of 4 Times Square.

34. William McDonough, the former dean of the architecture school at the University of Virginia, is noted for incorporating elements like rooftop lawns into buildings he has developed or redeveloped, such as Ford Motor Company's giant River Rouge complex in Dearborn, Michigan. But he most emphasizes the use of novel, environmentally friendly materials: see the book he wrote with Michael Braungart, *Cradle to Grave: Remaking the Way We Make Things.*

35. Stan Gale, the developer of New Songdo, confirmed in a message to the author on May 24, 2005, that green guidelines will be followed. The nonprofit U.S. Green Building Council formulates guidelines that a number of states, including Michigan, Washington, and Arizona, have adopted as policy. Collectively the guidelines are known as Leadership in Energy and Environmental Design (LEED).

36. "Green Manhattan," *The New Yorker* (October 18, 2004):111.

37. Smil, *Energy at the Crossroads,* 326 and 334.

38. See the annual reports from the Worldwatch Institute, Washington, DC, called *Vital Signs*.

39. Cerf is quoted in Huber and Mills, *The Bottomless Well,* 120.

40. Smil, *Energy at the Crossroads,* 332.

41. Huber and Mills, *The Bottomless Well,* 114.

42. Smil, *Energy at the Crossroads,* 335.

43. Quoted in Huber and Mills, *The Bottomless Well,* 112–13.

10. NATURAL GAS, GASOLINE, AND THE
VISION OF A HYDROGEN ECONOMY

1. "The Liberation of the Environment," *Daedalus* 125 (3) (Summer 1996): 3.

2. UTC Fuel Cells, in South Windsor, Connecticut, is a subsidiary of United Technologies, Hartford.

3. FuelCell Energy, based in Danbury, Connecticut, has delivered close to 30 molten carbonate units worldwide, most of them rated at 250 kilowatts.

4. See Bill McKibben, "Crossing the Red Line," *The New York Review of Books,* June 10, 2004, 33.

5. Danny Hakim, "George Jetson, Meet the Sequel," *The New York Times,* January 9, 2005.

6. It's largely because of Ovshinsky's innovations in nickel-metal battery technology that hybrid-electric cars have taken off.

7. American Physical Society, Panel on Public Affairs, "The Hydrogen Initiative," March 2004. This report can be downloaded from PoPA's home page, http://www.aps.org/public_affairs/index.cfm.

8. National Academy of Sciences, *The Hydrogen Economy.* See also Donald Anthrop, "Hydrogen's Empty Environmental Promise," briefing paper, The Cato Institute, Washington, DC, December 7, 2004.

9. Huberts was quoted in Robert F. Service, "The Hydrogen Backlash," *Science* 305 (August 13, 2004): 959.

10. Battaglia was quoted in "Scientist, Cartoonist Both Take on Hydrogen Plan," *The Electricity Journal* 17 (7) (August 2004): 6.

11. *Science* 305 (August 13, 2004): 917.

12. The Set America Free manifesto, which in March 2005 was resent to President Bush as an open letter, can be accessed at www.iags.org/safn.pdf

13. See Rifkin, *The Hydrogen Economy: The Next Great Economic Revolution* (New York: Tarcher/Penguin, 2002), and "Amory B. Lovins's Hydrogen Primer: A Few Basics About Hydrogen," http://www.rmi.org/sitepages/art75167.php.

14. The author participated in a couple of the events; for a sardonic description and analysis of the exercise, see Elizabeth Kolbert, "The Car of Tomorrow," *The New Yorker* (August 8, 2003): 36–40.

15. *The Wall Street Journal,* October 13, 2004.

16. See Glenn Zorpette, "The Smart Hybrid," *IEEE Spectrum* 41 (1) (January 2004).

17. See Willie D. Jones, "Take That Car and Plug It," *IEEE Spectrum* 42 (7) (July 2005), and Danny Hakim, "Hybrid Car Tinkerers Scoff at No-Plug-In Rule," *The New York Times*, April 1, 2005.

18. See Smil, *Energy at the Crossroads*, 207.

19. See Molly Espey, "Gasoline Demand Revisited: An International Meta-Analysis of Elasticities," *Energy Economics* 20 (3) (June 1, 1998): 273–95; Hilke A. Kayser, "Gasoline Demand and Car Choice: Estimating Gasoline Demand Using Household Information," *Energy Economics* 22 (3) (June 1, 2000): 331–48; and Steven L. Puller and Lorna A. Greening, "Household Adjustment to Gasoline Price Change," *Energy Economics* 21 (1) (February 1, 1999): 37–52. I am very grateful to Michael Ash at the University of Massachusetts, Northampton, for drawing my attention to this literature.

20. Joseph J. Romm, "The Hype About Hydrogen," *Issues in Science and Technology* 20 (1) (Spring 2004): 80.

21. See William Sweet, "Energy," in Foreign Policy Association, *Great Decisions* (New York, 2006), chapter 3.

11. A SECOND LOOK AT NUCLEAR ENERGY

1. See the reference to the University of Chicago study below.

2. Still relevant is Nordhaus, *The Swedish Nuclear Dilemma*.

3. For a nontechnical account of the accident, based on extensive interviews with leading nuclear engineering experts and close reading of the technical literature, see William Sweet, "Chernobyl: What Really Happened," *Technology Review* 92 (5) (July 1989): 42–52. For a somewhat more technical rendering, see John Ahearne, "Nuclear Power After Chernobyl," *Science* 236 (May 8, 1987): 673–79.

4. Of the many exhaustive technical reviews of what happened at Chernobyl, the most important is the Soviet Union's own report, and especially figure 4, which showed that the reactor's power level in the second explosion increased by a factor of nearly 50,000 above its rated maximum power level. See USSR State Committee on the Utilization of Atomic Energy, *The Accident at Chernobyl Nuclear Power Plant and Its Consequences*, information compiled for the IAEA experts meeting, August 25–29, 1986, Vienna. The main conclusions and lines of analysis were largely confirmed by other meticulous reviews, including those undertaken by the U.S. Department of Energy, the U.S. Nuclear Regulatory Commission, Canada's Atomic Energy Board, the International Atomic Energy Agency, and the Organization for Economic Cooperation and Development.

5. Christopher Flavin, "Reassessing Nuclear Power: The Fallout from Chernobyl," report for Worldwatch Institute, 1987.

6. Grigori Medvedev, a Soviet nuclear engineer, describing the explosions, refers to the "destruction, melting and then evaporation of the nuclear fuel"; he estimates that 50 tons of fuel vaporized in what he considers a "nuclear explosion," and that another 70 tons were ejected in solid form, sideways from the reactor. The

total reactivity of the ejected fuel was 15,000 to 20,000 roentgens (an international unit of electrostatic charge produced by x-radiation or gamma radiation) per hour. See his *The Truth About Chernobyl*, 78. Some experts, including John Ahearne and Zhores Medvedev, believe that the second explosion may have been a hydrogen explosion, but the more straightforward and simpler explanation is that it was a second reactivity excursion, as the Soviet simulation strongly suggests. For a statement of this view, see the letter to *The Bulletin of the Atomic Scientists* ([May–June 1997]:3 and 59–60) by George S. Stanford of Argonne National Laboratory. For Z. Medvedev's views, see *The Legacy of Chernobyl*, 26–33. (This Medvedev is best known in the West for a *samizdat* account he wrote during the Soviet period of a catastrophic explosion at a Russian nuclear waste facility, which of course had been kept secret.)

7. The light-water reactor is so named because its nuclear reactions are moderated by regular water rather than by heavy water, which consists of deuterium and oxygen. The Canadian heavy-water reactor, the CANDU, is in fact susceptible to the same kind of self-escalating nuclear excursion that occurred in the Chernobyl reactor, but the country's regulatory officials assured this author and Canadian parliamentarians that its control rods are fast enough to nip any such incident in the bud.

8. The Chernobyl Forum, *Chernobyl's Legacy: Health, Environmental and Socio-economic Impacts* (twenty-year assessment by IAEA, WHO, FAO, etc.) (Vienna: IAEA, September 2005).

9. The Nobel Prize–winning economist Gary Becker, however, has asserted that nuclear costs have come down and fuel prices for coal, gas, and oil have gone up so much that nuclear is now price-competitive. See "The Nuclear Option," *The Wall Street Journal*, May 12, 2005, editorial page.

10. As discussed in the University of Chicago study "The Economic Future of Nuclear Power," by George S. Tolley et al., August 2004. Selecting a lower discount rate admittedly seems contrived. But with the procedures used in the Chicago study and any other relying on the same cost-benefit methodology, the costs of nuclear power turn out to be almost completely insensitive to whether plants last just 40 years, 60 years, or even longer. This is because after several decades, given a discount rate of 8 or 10 percent, the long-term benefits are reduced to zero; that is, even though a plant is fully paid for and producing carbon-free electricity at a very low cost, those benefits count for nothing when the energy is discounted to present value. Because of this seemingly bizarre outcome, some argue for arbitrarily using a lower discount rate.

11. John Deutch and Ernest Moniz, "Nuclear Power Can Work," *The New York Times*, August 14, 2003, op-ed page.

12. University of Chicago, Tolley, abstract, p. x.

13. For a description of the main ones, still reasonably current, see William Sweet, "Advanced Reactor Designs," *IEEE Spectrum* 35 (4) (November 1997).

14. Several utilities and independent power producers, including Duke Power, Exelon, and Entergy, are at this writing seriously considering seeking permission to build the next new U.S. nuclear power plant.

15. Exelon and Entergy were doing better than the corporate average on the stock market in 2004 and 2005.

16. The author visited Indian Point with Philip Schewe of the American Institute of Physics on November 8, 2004.

17. See National Research Council, "The Safety and Security of Commercial Spent Nuclear Fuel Storage: Public Report," (Washington, DC: National Research Council, 2005); also, the *New York Times* editorial on the same subject, April 9, 2005. The academy report recommended two simple ways of significantly improving security at the more vulnerable ponds.

18. The superiority of dry-cask storage is addressed both in the academy report and, by implication, in "Nuclear Power and Proliferation Resistance: Securing Benefits, Limiting Risk," Nuclear Energy Study Group for the American Physical Society Panel on Public Affairs, May 2005. For a discussion, see Prachi Patel Predd, "Perils of Plutonium," *IEEE Spectrum* 42 (7) (July 2005).

19. See Matthew L. Wald, "A New Vision for Nuclear Waste," *Scientific American* 292 (3) (March 2003): 60–69.

20. For some alternative assessments of Yucca Mountain, see Jeff Wheelwright, "Welcome to Yucca Mountain," *Discover* 23 (9) (September 2002): 67–75, and B. John Garrick vs. Victor Gilinsky, "Yucca Mountain Pro and Con," *IEEE Spectrum* 39 (10) (October 2002).

21. John Kerry's stand against Yucca Mountain was criticized, for example, by Columbia University's James Hansen, for its inconsistency with Kerry's concerns about global warming.

22. See Nuclear Energy Study Group, American Physical Society Panel on Public Affairs, "Nuclear Power and Proliferation Resistance," 20.

23. In a breeder, an inner fuel core consisting of virtually pure fissile uranium and plutonium (U-235 and Pu-239) is surrounded by a "blanket" of nonfissile U-238. The very intense and fast reactions in the core produce a dense stream of neutrons, which are captured by the U-238, converting it to fissile Pu-239. In contrast to conventional reactors that run on a very dilute mixture of fissile and nonfissile uranium, a true or, if you will, "fast" nuclear explosion can occur in a breeder—the sanitized term preferred by the industry is "prompt critical burst."

24. See William Sweet, "Iran's Nuclear Program Reaches a Critical Juncture," *IEEE Spectrum* 41 (6) (June 2004), http://www.spectrum.ieee.org/WEBONLY/resource/jun04/0604niran.html#sb1.

25. Harold A. Feiveson, "Nuclear Power, Nuclear Proliferation, and Global Warming," Forum on Physics & Society of the American Physical Society, January 2003, *http://www.aps.org/units/fps/jan03/article3/html;* also, Feiveson, "The Dilemma of Nuclear Power," Symposium on the Role of Nuclear Power, University of Michigan, October 2–4, 2002, available from author.

26. Great Britain, Germany, and Spain all have ambitious plans to expand wind capacity, as noted in chapter 9.

27. Israel acquired a reactor from France in the 1950s with the specific intention of developing nuclear weapons. By the early 1960s it was obvious to the U.S. government that Israel was doing so, and a debate ensued in the White House as to whether the United States should try to stop this. It was decided in the end to turn a blind eye.

28. When Iraq acquired a reactor from France somewhat similar to Israel's, its intentions were so obvious that first, Israeli agents blew the reactor up before it was shipped from France, and then, when a second one was shipped and installed, Israel sent in F-16s to bomb it. After the first Gulf War, it was discovered that Saddam Hussein had an extensive nuclear weapons development program, including a · large uranium enrichment effort based on World War II technology.

29. It's true that when India tested what it called a "peaceful nuclear device" in 1974, it made some use of nuclear energy technology supplied by Canada. But from the time the international nuclear safeguards system was negotiated in the mid-1950s to the conclusion of the Nuclear Nonproliferation Treaty in the 1970s, India made it emphatically clear that it considered illegitimate the division of the world into nuclear weapons and non-nuclear weapons states. The underlying message was unmistakable: either the "nuclear club," consisting of the five World War II victor nations, would be abolished (that is, the nuclear weapons states would agree to undertake complete and general nuclear disarmament) *and* India would be accorded similar great power status; or India would acquire nuclear weapons, which its leaders considered their "ticket to the table." Two decades later, having been ignored, India went nuclear.

30. When India crossed the nuclear weapons threshold, Pakistan immediately followed, its leader having promised after the 1974 Indian test that it would acquire nuclear weapons even if its citizens had to "eat grass." Key to Pakistan's program was the theft during the 1960s of European enrichment technology from a plant in the Netherlands, which became the germ of the giant black-market nuclear technology network operated by the Pakistani nuclear scientist A. Q. Khan.

31. The apartheid regime, a pariah state, developed nuclear weapons as a hedge against foreign intervention or invasion by neighboring African states and guerrilla movements; upon agreeing to end apartheid, the minority white government quietly abolished the nuclear weapons program under international supervision, evidently so as to not turn it over to the incoming black majority regime.

32. Top U.S. diplomats, for example, went out of their way to persuade dubious customers like Ferdinand Marcos of the Philippines to place large nuclear orders with U.S. suppliers.

33. Weingart, *Waste Is a Terrible Thing to Mind.* See also the chapter about Weingart in Hall, *Real Lives, Half Lives*, 117–21.

34. Interview with the author, April 18, 2005; in the interest of fair disclosure, the author is a former classmate and an admirer—but not a personal friend—of Weingart.

35. Richard A. Meserve, "Global Warming and Nuclear Power," *Science* 303 (January 23, 2004): 433.

36. National Energy Policy Initiative, *Expert Group Report* (2003–04), created and coadministered by the Rocky Mountain Institute and the Consensus Building Institute, with the support of eight foundations; and the National Commission on Energy Policy, "Ending the Energy Stalemate: A Bipartisan Strategy to Address America's Energy Challenges," December 2004.

37. See Nicholas D Kristof, "Nukes Are Green," *The New York Times*, April 9, 2005, op-ed page, and April 12, 2005, letters in response.

38. See Thomas Homer-Dixon and S. Julio Friedmann, "Coal in a Nice Shade of Green," *The New York Times*, March 25, 2005, op-ed page.

39. Peter Schwartz and Spencer Reiss, *Wired*, February 2005.

CONCLUSION

1. Herodotus, *The History*, trans. David Grene (Chicago: University of Chicago Press, 1987), 55, 73.

2. T. P. Barnett et al., "Potential Impacts of a Warming Climate on Water Availability in Snow-Dominated Regions," *Nature* 438 (November 17, 2005): 303.

3. Purdue University press release, October 17, 2005.

4. The team, which made use of Germany's highest-performing computers and based its forecast on a model that had been tested against real-world climate developments of the last hundred years, reported its findings to the IPCC. Reported in *Terra Daily*, October 7, 2005.

5. P.C.D. Milly et al., "Global Pattern of Trends in Streamflow and Water Availability in a Changing Climate," *Nature* 438 (November 17, 2005): 347.

6. As the British barriers have had to be closed ever more often in recent years, the British government decided in 2004–05 to reinforce them. See Justin Mullins, "London Broil," *IEEE Spectrum* 42 (3) (March 2005).

7. See Nils Leffler, "Emissions Trading," *ABB Review* (March 2005): 14–19; Mark Ingebretsen and William Sweet, "Emission Permission," *IEEE Spectrum* 40 (1) (January 2003).

BIBLIOGRAPHY

Alley, Richard B. *The Two-Mile Time Machine: Abrupt Climate Change and Our Future.* Princeton, NJ: Princeton University Press, 2000.

American Institute of Physics, History Center. "Simple Models of Climate." www. aip.org/history/climate.

——. "General Circulation Models of the Atmosphere." www.aip.org/history/climate.

——. "GCM Historical Overview." www.aip-org/history/sloan/gcm/histoverview/html.

American Physical Society, Panel on Public Affairs. "The Hydrogen Initiative." College Park, MD: American Physical Society, March 2004.

Broecker, Wallace S. *The Global World According to Wally.* 3rd ed. Palisades, NY: Eldigio Press, 2002.

Ceruzzi, Paul E. *A History of Modern Computing.* Cambridge, MA: MIT Press, 1999.

Christianson, Gale E. *Greenhouse: The 200-Year Story of Global Warming.* New York: Penguin, 1999.

Dansgaard, Willi. *Frozen Annals: Greenland Ice Sheet Research.* Copenhagen: Aage V. Jensens Fonde, 2004.

Davis, Devra. *When Smoke Ran Like Water: Tales of Environmental Deception and the Battle Against Pollution.* Cambridge, MA: Perseus, 2002.

Deutch, John M. and Richard K. Lester. *Making Technology Work: Applications in Energy and the Environment.* Cambridge, UK: Cambridge University Press, 2004.

Diamond, Jared. *Collapse: How Societies Choose to Fail or Succeed.* New York: Viking, 2004.

Fagan, Brian. *The Little Ice Age: How Climate Made History, 1300–1850.* Cambridge, MA: Perseus, 2000.

——. *Floods, Famines and Emperors: El Niño and the Collapse of Civilization.* New York: Basic Books, 1999.

Freese, Barbara. *Coal: A Human History.* Cambridge, MA: Perseus, 2003.

Gelbspan, Ross. *The Heat Is On: The High Stakes Battle Over Earth's Threatened Climate.* Cambridge, MA: Perseus, 1997.

Gissen, David, ed. *Big and Green: Toward Sustainable Architecture in the Twenty-first Century.* Washington, DC: National Building Museum, 2002.

Gleick, James. *Chaos: Making of a New Science.* New York: Penguin, 1987.

Gore, Albert. *Earth in the Balance: Ecology and the Human Spirit.* New York: Penguin Plume, 1993.

Hall, Jeremy. *Real Lives, Half Lives.* New York: Penguin, 1996.

Hallam, A. *Great Geological Controversies.* Oxford, UK: Oxford University Press, 1983.

Houghton, Sir John. *Global Warming: The Complete Briefing.* 3rd ed. Cambridge, UK: Cambridge University Press, 2004.

Huber, Peter and Mark Mills. *The Bottomless Well: The Twilight of Fuel, the Virtue of Waste, and Why We Will Never Run Out of Energy.* New York: Basic Books, 2005.

Johannsen, Kristin, Bobbie Ann Mason, and Mary Ann Taylor-Hall, eds., *Missing Mountains.* Nicholasville, KY: Winding Publications, 2006.

Lomborg, Bjørn. *The Skeptical Environmentalist: Measuring the Real State of the World.* Cambridge, UK: Cambridge University Press, 2001.

Lyman, Francesca, et al. *The Greenhouse Trap: What We're Doing to the Atmosphere and How We Can Slow Global Warming.* Boston: Beacon Press, 1990.

McDonough, William and Michael Braungart. *Cradle to Grave: Remaking the Way We Make Things.* New York: North Point Press, 2002.

McPhee, John. *Annals of the Former World.* New York: Farrar, Straus & Giroux, 1998.

Medvedev, Grigori. *The Truth About Chernobyl.* New York: Basic Books, 1991.

Medvedev, Zhores. *The Legacy of Chernobyl.* New York: Norton, 1990.

Nash, J. Madeleine. *El Niño: Unlocking the Secrets of the Master Weather Maker.* New York: Warner, 2002.

National Academy of Sciences. *The Hydrogen Economy: Opportunities, Costs, Barriers and R & D Needs.* Washington, DC: National Academy of Sciences, 2004.

National Research Council. *Novel Approaches to Carbon Management: Separation, Capture, Sequestration, and Conversion to Useful Products.* Washington, DC: National Academies Press, 2003.

——. *Abrupt Climate Change: Inevitable Surprises.* Washington, DC: National Academies Press, 2002.

——. *The Carbon Dioxide Dilemma: Promising Technologies and Policies.* Washington, DC: National Academies Press, 2002.

Nebeker, Frederik. *Calculating the Weather: Meteorology in the Twentieth Century.* New York: Academic Press, 1995.

Nordhaus, William D. *The Swedish Nuclear Dilemma: Energy and the Environment.* Washington, DC: Resources for the Future, 1997.

Posner, Richard A. *Catastrophe: Risk and Response.* Oxford, UK: Oxford University Press, 2004.

Reece, Erik. *Lost Mountain: Radical Strip Mining and the Devastation of Appalachia.* New York: Riverhead Books, 2006.

Robinson, Walter A. *Modeling Dynamic Climate Systems*. New York: Springer-Verlag, 2001.

Ruddiman, William. *Earth's Climate: Past and Future*. New York: Freeman, 2001.

Schwartz, Peter and Doug Randall. "An Abrupt Climate Change Scenario and Its Implications for United Sates National Security." Washington, DC: U.S. Department of Defense, October 2003.

Sheets, Bob and Jack Williams. *Hurricane Watch: Forecasting the Deadliest Storms on Earth*. New York: Vintage, 2001.

Smil, Vaclav. *Energy at the Crossroads: Global Perspectives and Uncertainties*. Cambridge, MA: MIT Press, 2003.

Speth, James Gustave. *Red Sky at Morning: America and the Crisis of the Global Environment*. New Haven: Yale University Press, 2004.

Weart, Spencer R. *The Discovery of Global Warming*. Cambridge, MA: Harvard University Press, 2003.

Weingart, John. *Waste Is a Terrible Thing to Mind: Risk, Radiation and Distrust of Government*. Princeton, NJ: Center for Analysis of Public Issues, 2001.

Winchester, Simon. *The Map That Changed the World*. New York: HarperCollins, 2002.